Science, Technology, and Society

Science, Technology, and Society

A Sociological Approach

Wenda K. Bauchspies,
Jennifer Croissant,
and Sal Restivo

Blackwell
Publishing

BLACKWELL PUBLISHING
350 Main Street, Malden, MA 02148-5020, USA
9600 Garsington Road, Oxford OX4 2DQ, UK
550 Swanston Street, Carlton, Victoria 3053, Australia

First published 2006 by Blackwell Publishing Ltd

1   2006

Library of Congress Cataloging-in-Publication Data

Bauchspies, Wenda K.
Science, technology, and society : a sociological approach / Wenda K.
Bauchspies, Jennifer Croissant, and Sal Restivo.
p. cm.
Includes bibliographical references and index.
ISBN-13: 978-0-631-23209-4 (acid-free paper)
ISBN-10: 0-631-23209-5 (acid-free paper)
ISBN-13: 978-0-631-23210-0 (pbk. : acid-free paper)
ISBN-10: 0-631-23210-9 (pbk. : acid-free paper)
1. Science—Social aspects. 2. Technology—Social aspects. 3. Technology—
Sociological aspects. I. Croissant, Jennifer, 1965– II. Restivo, Sal P. III. Title.

Q175.5.B384 2006
303.48′3—dc22
2005002632

A catalogue record for this title is available from the British Library.

Set in 10/12.5pt Dante
by Graphicraft Ltd., Hong Kong
Printed and bound in India
by Replika Press Pvt. Ltd.

The publisher's policy is to use permanent paper from mills that operate a
sustainable forestry policy, and which has been manufactured from pulp
processed using acid-free and elementary chlorine-free practices. Furthermore,
the publisher ensures that the text paper and cover board used have met
acceptable environmental accreditation standards.

For further information on
Blackwell Publishing, visit our website:
www.blackwellpublishing.com

Contents

Preface   vii

1   Introduction   1
Our Premises and Our Approach   2
Nuts and Bolts and the Systems that Tighten Them   5
From the Sociology of Science to Science and Technology Studies   11
Conclusion   16

2   Cultures of Science   19
Birth of a Fact   19
Society and Culture   21
Worldviews   23
The Social Construction Conjecture   23
Feminism and Science Studies   26
Technology in Motion   32
Pre-scientific: You or Me?   36
Mind and Society   41
What Can Sociologists Say about Mathematics?   44
Conclusion   46

3   The Dance of Truth   49
Science and Technology as Social Institutions   51
The Dance of Magic, Science, and Religion   54

What Is Truth?		62
Dangerous Icons: From Magic and Religion to Science and Law		67
Conclusion		70

4 STS and Power in the Postmodern World — 73
Technology and Society — 73
Power, Values, and Agency — 85
Cyborgs, Humans, and Technology — 90
Contemporary Society: Globalization or Bust? — 93
Metaphors, Narratives, and Glocal Cultures — 97
Conclusion: Technoscience and Globalizations — 100

5 Life after Science and Technology Studies — 102
Technoscience Revisited — 102
Case Study: The New Reproductive Technologies — 104
Case Study: Robots, Minds, and Society — 107
Frontiers and Horizons — 110
Conclusion: Where We Have to Stand in Order to Begin — 115

Glossary — 121
References — 129
Name Index — 142
Subject Index — 146

Preface

Sc   w field of study,
in   aracteristics of a
discipline or field. An origin story could identify such key moments in the birth process of this field as the first publication of *Science Studies* (later *Social Studies of Science*) in Edinburgh in 1971, the founding of the Society for Social Studies of Science in 1975, and the historical first meeting of the society at Cornell University in 1976. It is clear that STS is something different from its constitutive elements – history, philosophy, and sociology of science and technology. It is not so clear what STS is except that it is a hybrid discipline and interdisciplinary. Therefore, its practitioners are still looking for a relatively stable identity. There are many efforts being undertaken to establish its character, from introductory texts, such as this one, to workshops/seminars at conferences and universities, conversations in journals, newsletters and listservs, training of graduate students, and creation of handbooks and encyclopedias. From these forms of dialogue, it is also evident that at least some practitioners are resistant to attempts to solidify STS as a field. One thing is clear: an identity for STS is, and must be, based upon pluralities and diversities, and their resulting dialogue(s). We are thus on the side of those practitioners who strive to sustain the hybrid, interdisciplinary, noncanonical state of the field rather than to seek or force unification. At the same time, we believe the field is more coherent than some of our colleagues claim, with a central dogma

(the social nature of science and technology) and a core literature that if not canonical nonetheless serves to define the field.

STS has some fundamentally disciplinary interests and objectives. The STS textbooks that have appeared have been efforts, such as this one, to develop a primer for the field.[1] However, they have not followed the familiar outline that one expects in comparable textbooks from other disciplines. Their chapter outlines do not, like chapter outlines in competing disciplinary textbooks, look pretty much alike. They have developed idiosyncratic stories about the field and our text is no exception. We reflexively understand that we are doing our own social construction of a creation story and canon. What recommends the book in the first place is our dedication to STS as our intellectual home. Second, we are committed to the idea that STS is one of the sociological sciences. If there is a central dogma in STS, it is that science, technology, knowledge, and belief are social constructions, or to put it more mildly and to make it more palatable to more of our colleagues: science and technology, or the technosciences, are social and cultural phenomena. The emergence and development of STS is centered in the social sciences. Our bias in this book is that we will emphasize the sociological sciences as the foundation for STS and spend some effort outlining the intellectual trajectory that brought us to the point of writing this book.

Finally, one of the things we would like to accomplish in this book is to make it as clear as possible that science and technology use the tools of the social sciences and humanities to study, understand and analyze science, technology and the work of engineers and scientists past and present. Unfortunately, many have misunderstood our theoretical stance as aggressive rather than explanatory, or as an attack on truth, objectivity, or reality. It is understandable that in our more critical approaches we may ruffle the feathers of scientists, science worshippers, and science watchers. Their counter-attacks, however, often extend to the basic explanatory objectives that are the foundation of STS. While we may be critical of modern science as a social institution, our theoretical position is not based on denying an antecedent reality. What we do deny is the idea that there is an already and always existing description of reality that we approach through closer and closer approximations.

We are not ready to impose nor are we interested in imposing upon the field or reader a strict and monolithic view of STS. This book is about our experiences in and with the field. We would like it to stand not as an absolute alternative to competing STS introductions but as a simple alternative, maybe even one to be read side by side with other texts and readers. Furthermore, while we introduce case studies, we would encourage readers to consult as collateral readings books that review science studies from a more focused empirical perspective, such as the two books by Harry Collins and Trevor Pinch (1998a, 1998b), or the numerous case materials by our colleagues, listed in the bibliography.

Who Are We?

We are social theorists. We work, read, and do research under various disciplinary labels from sociology and cultural anthropology to political science and philosophy. Everyone is a social theorist; the fact that you are reading this text is an example of your expertise at negotiating the social world of schools, learning, and reading. Very simply, a social theorist is what you are when you move through your everyday/everynight world figuring out what others are doing around you, with you and to you. Social theory is a basic survival skill that you turn to in order to buy food, go to class, greet someone, and make a friend. This sort of everyday/night (folk) social theory is put in use without the self-consciousness of a professional social theorist. As professional social theorists, we name and talk about the hidden aspects of social life that are often labeled mundane and unimportant (Lemert 1993). We talk about the power dynamics of the classroom, the "neutrality" of technology and the "truth" of science in order to "ground activities previously seen as individual, mental and non-social as situated, collective and historically specific" (Bowker and Star 1999, 288). Understanding our world, our knowledge, and our artifacts as situated collectively and historically enables us to explore un-articulated ideas and concepts, challenge pre-existing notions, begin dialogues, make changes, shift power, and alter perspectives.

Once upon a time the first author, Wenda Bauchspies, taught physics. She enjoyed teaching students about how the physical world worked. However, she found students far more interested and successful in learning the content of science when it was placed within a socio-cultural and historical framework. That meant, for example, thinking about the development of the steam engine within the industrial revolution, learning chemistry by looking at pollution, and exploring cell biology by learning about nutrition. One thing led to another and Bauchspies found herself in a new interdisciplinary field called Science and Technology Studies (STS) where the third author, Sal Restivo, a founder of the field of STS, became her mentor and the second author, Jennifer L. Croissant, became her academic "big sister." Thus it is not an accident that the three of us have joined together to write this introductory STS textbook to help readers think about the interfaces of science, technology, and society. Our collaboration is an example of how social networks develop and create knowledge. We will discuss this further in chapter 2.

The second author, Jennifer Croissant, sometimes describes herself as a "lapsed engineer" because although she had engineering training as an undergraduate, she never worked as an engineer. Croissant chose not to take a job which would have involved designing missile vision systems and went on to graduate school in technology and public policy to figure out why the defense industry seemed to be about the only place jobs were available for engineers. In a public policy program, she learned to evaluate technological systems using cost–benefit analysis and other analytical tools. However, the analysis never included questions about whether or not a technology was justifiable on cultural or moral grounds, and whether or not it was worth pursuing at all. Thus, Croissant found her intellectual home in Science and Technology Studies where these sorts of questions are asked with the help of theories from history, sociology, anthropology, political science, and philosophy. For the last several years, she has taught in the general education program at the University of Arizona. Her contributions to this textbook emerge from course materials, lecture notes, and stories, which have been designed to get non-science students to think about science and technology, and to challenge the science, engineering, and other pre-professional students to think about their social responsibilities in their chosen professions. This is not always an easy task given that frequently our assumptions about how the world works are very different from those of our students and our science and engineering colleagues.

Sal Restivo had an early education that focused on science and mathematics, and later on electrical engineering. In the end, he came to sociology and anthropology because they spoke to a deep-seated need to understand himself and his world. In graduate school, he discovered that there was a specialization within sociology that would allow him to draw on his background in, familiarity with, and continuing interest in science and mathematics: the sociology of science. At the same time that Restivo was familiarizing himself with this field of study in sociology, the field itself was already undergoing a significant transformation, the birth of science and technology studies.

Now let us situate ourselves in a longer narrative of studying science and technology. In nineteenth-century Europe, at the same time that the social sciences were being fashioned in the molds constructed out of the social practices, ideologies, and mythologies of the physical sciences, the path was being cleared for a sociological theory of science. Early thinkers, including Peter Kropotkin and Karl Marx, recognized in the midst of fashioning human sciences that science was social relations. Writing in the early 1840s in England, watching the industrial revolution unfold, Marx noted in *The Economic and Philosophic Manuscripts of 1844* (1959, 104) that everything from our behaviors to our thoughts and our very selves was social:

> Even when I carry out scientific work, etc., an activity which I can seldom conduct in direct association with others – I perform a social, because human, act. It is not only the material of my activity – like the language itself which the thinker uses – which is given to me as a social product. My own existence is a social activity.

Here, then, we have in the space of a few lines the ideas that the self, the mind, language, and science are social constructions. Somewhat later, Max Weber described the cultural context of science, and the functions of material resources and social structures in the development of science – and he realized as acutely as Marx the connection between modern science and capitalism. Perhaps no one saw as clearly as Marx and Emile Durkheim the fact that knowledge and concepts were social constructions. It is from this premise, articulated more fully in the next chapter, that we proceed.

Introduction to the Book

Chapter 1 is an introduction to the culture of science and technology studies that focuses on the foundations and background of the field that have contributed to the current ideas and theories found in this book. We introduce the reader to our language game and the words we use from science, technology, and culture to technoscience and technosocial. It ends with a bit of history on the sociology of science and the meaning of social worlds and thinking.

Chapter 2 begins with a discussion of facts and how they function in science. We explore the development, creation, and establishment of facts as constructs of society and culture. We introduce the concept of worldviews to further explore the meaning of *fact*. Social construction is discussed in more detail in relation to fact, society, and science. An illustration of these concepts and how they get played out beyond the pages of the book is explored in the sections on feminism and science studies, technology, and pre- or proto-scientific activity. Scientific facts shape how we understand our world and ourselves and the section on mind and society looks at our understanding of mind in light of a sociological worldview. Mathematics is often thought to be pure and separate from social phenomena, but the final section introduces the idea of mathematics as a social construction.

Chapter 3 begins with a discussion of science as culture and then develops an understanding of technoscience as an institution by looking at its similarities and differences compared to magic and religion, and the law as institutions. Our premise is that social institutions are powerful, productive, and also limiting and dangerous. In addition, we introduce ideas about truth, knowledge, and objectivity that are not particularly new, but often under appreciated.

Chapter 4 provides a focus on technology and society through exploration of the ideas of Lewis Mumford, Ivan Illich, Karl Marx, and Langdon Winner on technological society, technological determinism, technological "fixes," and technological agency. The discussion then moves on to technological change, innovation, cultural convergences, the so-called "neutrality" of technology, and technological adoption. We contextualize these through a discussion of globalization and colonization. The chapter ends with a discussion of some of the new directions in technology studies, from risk analysis to analyses of power, identity, and gender.

Finally, chapter 5 revisits technoscience and the question of what it could mean to live in a technoscience–technosocial world. We have chosen cases that highlight the complexities and ambivalences of life in a technoscientific society: reproductive technologies and social robotics. The questions we raise here and the analytic tools we introduce are designed to help engage our world in ways that support diversity, sustainability, multiplicity, freedom, and creativity.

We want to emphasize that our objective in writing this book is to prepare a primer that speaks to the uninitiated in a language that they recognize. Occasionally, we stray from this strategy to bring the reader closer to the central problems and perspectives of our field. In general, however, we have tried to speak directly to the reader new to science and technology studies.

<hr>

Note

1 Other STS primers and resources are: *Science, Technology and Society* by Robert McGinn, Englewood Cliffs, Prentice Hall, 1991; *Society and Technological Change* (3rd edn.) by Rudi Volti, St. Martin's Press, New York, 1995; *Science and Technology in a Multicultural World: The Cultural Politics of Facts and Artifacts* by

David Hess, Columbia University Press, New York, 1995; *Science Studies: An Advanced Introduction* by David Hess, New York University Press, New York, 1997; *Technology and the Future* (8th edn.), edited by Albert Teich, Bedford/St. Martin's Press, New York, 2000; *The Golem at Large: What You Should Know about Technology* by Harry Collins and Trevor Pinch, Cambridge University Press, Cambridge, 1998; *The Golem: What You Should Know about Science* (2nd edn.), by Harry Collins and Trevor Pinch, Cambridge University Press, Cambridge, 1998; *Visions of STS: Counterpoints in Science, Technology, and Society Studies* edited by Stephen Cutcliffe and Carl Mitcham, SUNY, Albany, 2001; *Introduction to Science and Technology Studies*, by Sergio Sismondo, Blackwell, London, 2003; *Chasing Technoscience* edited by Don Ihde and Evan Selinger, Indiana University Press, Bloomington, 2003.

Further Reading

Fuller, S. (2004). *The Philosophy of Science and Technology Studies*. London, Taylor & Francis.

Hackett, E. J., Amsterdamska, O., Lynch, M., and Wajcman, J. (eds.) (2007). *New Handbook of Science and Technology*, Cambridge, MIT Press.

Hess, D. (1997). *Science Studies*. New York. New York University Press.

Jasanoff, S., Markle, G. E., Petersen, J. C., and Pinch, T. (eds.) (1995). *Handbook of Science and Technology Studies*. Thousand Oaks, Sage.

Restivo, S. (ed.) (2005). *Oxford Encyclopedia of Science, Technology, and Society*. New York, Oxford University Press.

Spiegel-Rosing, I. and de Solla Price, D. (eds.) (1977). *Science, Technology, and Society: A Cross-Disciplinary Perspective*. Thousand Oaks, Sage.

1

Introduction

Our objective is to share with you some of our curiosity about science and technology. Sociology, anthropology, philosophy, other social sciences and humanities disciplines have helped to fuel this curiosity because they provide us with a way to analyze, interpret, and understand science and technology as social relations and as socially constructed. STS is by no means a mainstream view of science and technology. Many find it threatening because it challenges traditional ways of thinking, experiencing, and responding to science and technology. Our goal here is to help you use the tools of social and cultural studies to study science and technology, to avoid the extremes of naïve realism and naive relativism, and to stay out of the "science wars." The science wars pit physical and natural scientists against the efforts of social scientists and humanities scholars to explain and theorize science and technology, and in some cases to subject it to social criticism. Often the STS scholar is denounced as being anti-science, anti-technology, or being naïve about science and technology. Many of the scientists and engineers who complain about sociologists of science do not understand sociology, and make unfounded assumptions about the discipline and its methods. We are not going to get involved in this debate here, but we do want to raise a caution flag for those of you who enter this field of study, and want to understand the interdisciplinary dialogues of STS. Interdisciplinarity is not easy and often requires careful listening and communicating by all because the conversation is occurring in a hybrid space.

Our Premises and Our Approach

We begin by introducing a certain way of talking and writing about science and technology that Ludwig Wittgenstein (1953/2001) describes as a language game. Languages come in different forms with different sets of rules, and they are embedded in different ways of living or forms of life. Language games entail placing these different language sets together to allow a more expansive and critical perspective to emerge. To give you experience with the STS language game, we are going to spend a few paragraphs demonstrating some of its basic features. We are going to deliberately be a little bit repetitive in the interest of saying things in different ways: these variations in description and definition provide different perspectives on the topics and concepts and highlight the contradictions and gaps that appear when only using a single or limited set of discourses. In addition this is also an opportunity for us to unfold the value orientations that guide our particular approaches to the study of science, technology, and society. Thus, the main feature of our language game is the use of the word *social*.

Humans are social beings. We begin there but we do not stop there. For it is not simply people who are social, it is also the worlds that they create, the ideas that they think and the artifacts they use. All are part of the social fabric within which we exist. We and our ideas and our artifacts are in the most fundamental way social through and through because all of these things exist in a web of social relationships. Thus, in this book we look at, interpret, and analyze society, individuals, ideas, and artifacts with social lenses. The reader may find some of our ideas strange, outrageous, counterintuitive, wrong, or even silly. Simultaneously, some of the ideas may articulate something the reader knows and has not yet been able to explain. We ask you to pay attention to the ideas that are either harmonious or dissonant for you, because these are important nodes that will further understanding. We encourage you to think about why these ideas are repulsive or comforting, or seem completely bizarre or totally appropriate. What do these ideas challenge or reinforce? How do these ideas work on the societal level and on the individual level? Or in other words what purpose do they serve, for whom, and why?

Some may want to say "stop" here and ask "if everything is social then where is truth, rationality, and objectivity?" You might say that you recently heard a scientist say that science cannot be a social construct because it is independent of society or deals strictly with the natural world. This is like claiming that humans cannot manufacture airplanes and still fit within natural systems of airflow and gravity. An airplane is obviously manufactured and functions because of our abilities to categorize, explain, and manipulate aspects of the natural world. It is not that facts and things are *either* socially constructed *or* true or reliable, but that they are *both* socially constructed *and* true or reliable, or perhaps false or flawed, depending on the circumstances. The issue here is that we as thoroughly social humans have not yet learned to understand and fully experience the social forces that surround us, that are us. Science and technology have allowed us to creatively engage the physical world and make airplanes. However, we have not collectively and as a society mastered understanding or using social science to the same extent. We are advocating the same attention to the social sciences that we as a society give to the physical and natural sciences. Our claim is that the social sciences are discovery sciences with powerful knowledge bases and can be applied to science itself.

We can claim that time, space, class, cause, and personality are socially constructed without denying their objectivity (Durkheim 1961, 31–32). The social construction of knowledge and science was a central theme in the ideas of the nineteenth and early twentieth century thinkers who crystallized the social sciences.[1] Oswald Spengler (1926, 100) describes science as the story that humans tell about themselves and characterizes the scientific experience as "spiritual self-knowledge." George Herbert Mead (1934, 186) equates being rational, reasonable and thoughtful as "taking a-social attitude toward the world about us."

As heirs to their achievements, we can now fearlessly broaden the dialogue to include the cognitive authority awarded to science and scientists, design of technology, long and short term policies, progress, efficiency, machines and so on. For if there is anything to fear, it is the unexamined exercise of any form of authority, cognitive or otherwise:

> Illegitimate politicization and rampant irrationality find their most fruitful soil when our activities are mystified and protected from criticism. (Addelson 1983, 182)

Like the giants whose shoulders we stand on, we are scientists; we have methods, communities of practice, education, and professionalization. We are scientists who study science and technology, and we want to share with you some of the tools and skills required to ask new and different questions about science, technology, society, and ultimately ourselves.

Throughout this text we ask important questions that break the boundaries across the disciplines. We not only want to know how society works, we want to understand the nature of the "good" society. We want to know how the organization of society affects us, and how we affect society. We also want to know what the roles of science and technology are in a good society. We want students to keep in mind that science and technology are social institutions that depend on social factors. This realization leads us to important insights and provocative questions. Who drives technological change? Who controls science? Why do we trust, and why do we disbelieve, various experts? How are science and technology linked to questions of values, ethics, and social justice?

We do our sociology without apologies: there is resistance to the word "social" and while we recognize the discomfort it causes, we also realize that discomfort signals an important area for discovery. We are not privileging sociology as a discipline, but privileging an epistemologically oriented worldview. We do this by paying attention to icons, metaphors, signs, and representations used by people to create their social worlds. Consider for example what sociologist Howard Becker (1982) does in his book, *Art Worlds*. He follows various actors engaged in the production of art and their practices in the art world. This close examination of the people involved and their activities shows us how certain kinds of objects and practices become "art" as they move through various art worlds. Becker doesn't argue foundationally about "what art is." We too are also not interested in arguing about "what science is." Rather, we prefer to ask about the whos, hows, whys, wheres, and whens. This, some might recognize, is informed by a pragmatist turn which is not so much interested in the logical or intellectual rationalization of what science is or should be, but in the exploration of concrete ways to study both science and nature with attention to the contexts and consequences of activities of inquiry, and without worshipping the symbols and icons produced by and about science.

While scientific and technological literacy are important to us, this book is oriented to improving sociological literacy: understanding how our social institutions work, and how the very fabric of our lives is social. This sociological literacy also feeds into ideas about critical literacy. Most scientific literacy has been focused on making sure that students know long lists of facts and figures about the physical and natural worlds, and are able to define and apply scientific concepts, and defer to scientific expertise. Most models of technological literacy focus on creating students who are good consumers and users of technologies. However, we wonder if students are good designers of technology? Can they evaluate technology based upon criteria other than price? Can they identify ethical issues surrounding a technology? Will they be able to help a loved one choose the best treatment for an illness? Can they make policy decisions about science and technology? Because answers to these questions are difficult, we would like students to understand science and technology within a context that includes culture, history, and values.

Thanks to new telecommunications technologies and the pervasive spread of traditional information technologies and media (books, newsprint, word-of-mouth), many of us have access to many kinds of information, and a great deal of it. Few people, however, understand how that information was acquired, how to assess its validity and reliability, and how to evaluate various standards of proof and legitimacy. In addition, it is important to understand the large-scale systems of our interconnected lives. We believe that it matters very much who pays for scientific research, what cultural assumptions provide the frameworks for reasoning, and what scientific networks are in action when scientific and technological statements and products appear in our own or other social worlds. We advocate a broader definition of "literacy" that includes how facts and figures come to be accepted, and how science and technology are developed and work in our lives and the lives of our neighbors.

Nuts and Bolts and the Systems that Tighten Them

To study science and technology sociologically we begin with thinking about what science and technology seem to be. We address what their stated purposes are in society, and what they symbolize. During the course of our discussion, we look at social institutions, symbols, power, and culture as tools for understanding science and technology.[2]

Science can be a rather slippery term. We recognize it as knowledge accumulated through "the scientific method," as statements of fact and theories or explanations for events based on procedures for testing knowledge. These methods of science are sometimes simply called "science." This distinction opposes formalized knowledge to craft or practical knowledge, revealed knowledge, intuition, common sense, and other forms of inquiry. Science is often understood as the primary source of knowledge in western societies since about the sixteenth century (CE).[3] We want to recognize that other cultures have ways of developing and transmitting valid knowledge, and often it is not organized in ways which contemporary natural philosophers recognize or approve. We define science for the moment as a social institution grounded in an explanatory strategy that does not have recourse to paranormal, supernatural, or transcendental causes. Sometimes the terms science and knowledge are used interchangeably: we want to emphasize that all societies produce knowledge, systems of beliefs and ideas about how the world works, and systems of practice that implement that knowledge.

A society is a group of people who share a culture, have economically interdependent members, and that (according to conventional definitions) reside in a specific geographical area. Society as defined by Emile Durkheim (1961) is dependent on people sharing "essential ideas," such as time, space, cause, and number. This sharing of concepts, even when their meanings are debated, gives individuals the categories that ground their communication. Karl Marx (1958, 104–105) offers some of the clearest statements in early social theory on the social nature of the individual. A good working definition for society is the collective sharing of concepts and cooperation by individuals in some manner to achieve a set of ends. Societies directly and indirectly affect and are affected by the activities of individuals.

Anthropologists study culture and describe it as "all that humans learn" (Jacob 1992) and "relationships between people" (Weiner 1976). An early anthropologist, Edward Tylor (1871/1958), stressed the capabilities and habits that are needed by an individual to belong to a society, such as knowledge, art, law, customs, morals, and belief. Another anthropologist, Ruth Benedict (1934), said that culture is the ideas and "standards that people share in common." One of the most commonly quoted definitions of culture is "a web of significance within a group or society, that is a public creation that controls and completes the individual" (Geertz 1973, 5). A culture is the shared values, beliefs, materials, and practices of a social group. It includes relationships, ideas, values, standards, boundaries, classification systems, communication, learning, social networks, and social contexts.

Another word that needs to be considered carefully is *social*. For example, the term "social," as in "social construction," is not a synonym for "political," "religious," "economic," or "ideological;" nor does it connote or denote "false" or "arbitrary." To say that facts (scientific or otherwise) are socially constructed is not to say that they are false, arbitrary, fabricated out of thin air, or the direct causal product of "external" political, religious, economic, or ideological forces. The original laboratory studies by Bruno Latour and Steve Woolgar (1979), Karin Knorr-Cetina (1981), Michael Zenzen and Sal Restivo (1982), and Sharon Traweek (1988) (today, we would call them ethnographies of science), for example, helped to document the moment-to-moment, day-to-day, night-to-night minutiae of social interactions that make up the social processes and institutions of invention and discovery. The social is not only in the "external" social and cultural milieu or context of science, but in the social organization of science, indeed in scientists themselves. The social in this sense is pervasive, and no more or less transparent than quantum or gravitational forces.

Technology, like science, is also a slippery term. The three most common meanings of the word technology are: physical objects, activities or processes, and "know-how" (Bijker, Hughes and Pinch 1987). The meaning is dependent on the context and objective of the author. Stephen Unger (1994, 3) discusses ethics in engineering and defines technology as "the intelligent organization and manipulation of materials for useful purposes." Unger's definition complements and enhances the thesis of his book that engineers are responsible for the technology they create. A very simple and elegant description of technology is "what we make and what we do" (Winner 1977, 9). In this discussion, there is no one meaning for technology, but its meaning is a summation of artifact, process, and knowledge gained from experience.

Often the distinction between science and technology is said to be like that between basic and applied knowledge or, in other words, that technology is applied knowledge while science is basic knowledge. However, careful historical study reveals that all knowledge is applied. Thus the question turns to communities of practice, with attention to who makes what distinction, where, and why. For example, new knowledge produced in physics might be applied to problems only of interest to theorists and thus be labeled "basic," while more "applied" work is relevant to a larger network of communities of practice that might include health professionals, manufacturers or materials science engineers. For a contemporary example of the convergence of knowledge and practice, consider the biotechnology industry, where fundamental processes of genetics and protein chemistry are explored with the goal of developing therapeutic products. Is biotechnology science? Technology? Or both? A new word, technoscience, entered our language toward the end of the twentieth century to help us answer these questions and more.

The term, *technoscience*, introduced by Bruno Latour (1987, 174), describes "all the elements tied to the scientific contents no matter how dirty, unexpected or foreign they seem." Latour uses the expression "science and technology," in quotation marks, to designate *"what is kept of technoscience once all the trials of responsibility have been settled"* (emphasis in the original). In other words what we name as science or technology is the "clean" artifact, idea, concept, law or theory. Messiness, ambiguity, gray areas are hidden once a community of practice labels something science or technology.

Another social theorist, Donna Haraway (1997, 50–51), uses the term technoscience to point to "a condensation in space and time, a speeding up and concentrating of effects in the webs of knowledge and power . . . In short, technoscience is about worldly, materialized, signifying and significant power." The terms science and technology collapse cultural practices, historical influences, and social relationships into a package. Technoscience carries power within its culture; this power is both placed upon and embedded in it.[4]

By using the word technoscience, we emphasize the "dirt," and the muddle of the scientific and the social. Or in other words technoscience emphasizes the opposite of science as "pure and abstract" and technology as "neutral." It is in technoscience that we can locate the loose ends, the gray areas, the failures, and the works in progress that at the end of the day will be named "science and technology." We focus on the gray areas to highlight socio-cultural webs surrounding the transfer of knowledge and objects. By thinking of science and technology as technoscience with all the messiness, the boundaries, the transfer and appropriation of individuals, technologies, and knowledge across cultures, we alter the static interpretation of science and technology as separate from society, culture, and social worlds. Properly understood, technoscience changes our understanding of "pure" and "abstract" and "neutral" when applied to science and technology. We begin to understand that something is labeled "pure" once it has gone through processes that abstract and reify knowledge, stripping away evidence of its social origins. The processes clean and press the messiness and dirt out of science and technology. This creates an image of seamlessness, ethereality, and uniqueness in the ideas, concepts, and artifacts of science and technology.

The "ironing out" of a fact includes material practices, such as purification of samples, isolation of systems, selection of processing materials, the training of technicians so they are reliable, or even the special breeding of standardized laboratory animals. It also includes an interesting set of textual transformations, whereby the representation of an observation by an actor becomes transformed into independent fact. For example, "I think I saw X," is depersonalized to "X was observed by the researcher." This is further edited to "It has been observed that X," and ultimately transformed to "X is."[5]

Besides erasing the observer by developing a textual objectification, other strategies include erasing the work, and workers, that support technoscience. Technicians and other people become invisible in science in part because of their subordinate status, and often compounded by their social positions, such as race, class, gender. For example, Vivien Thomas, an African–American man, was a medical laboratory technician from 1929 to 1979. Dr. Blalock, a surgeon, trained Thomas at Vanderbilt and invited Thomas to accompany him to Johns Hopkins University. During his career as Blalock's technician, Thomas performed laboratory research and developed new surgical techniques at a time when blacks who worked for the university were listed as janitors in the account book.

> In a workplace where technical skills are highly valued, the laboratory offered Thomas a refuge for his inventiveness and dexterity. But it also generated misalignment, dissonance, and personal suffering when his work accumulated credit and prestige for others but not for him. The torque of institutionalized racial inequality was exacerbated by the twisting forces of great accomplishments and belated credit. (Timmermans 2003, 220)

Thomas's history and the cultural moment in which he lived and worked highlights the erasure then, and rewriting now, that occurs when we study the practices of technoscience and attend to race and hierarchy. It raises questions about who gets credit for what, who gets written into, and out of, the stories of scientific production. Unfortunately Vivien Thomas's story is not unique in science and the continual subordination of invisible workers contributes to the illusion of objectivity, democracy, and truth in science while hiding its discrimination, violence, and power.[6]

Once something is labeled "pure," it symbolizes that it is free of entanglements and relationships. Labeling something "science" in essence tends to declare its purity and ignore all of the antecedent work that went into its construction. We can already see in fields like biotechnology that the distinctions between pure and applied science and between science and technology are no longer viable. As we look anew at what we have called science and technology in the past, it turns out that those distinctions have not been as clear as we always believed. By the time you finish this book, you may begin to see that science and technology are abstractions from a reality in which there is only technoscience.

We will sometimes use the word *technosocial* to direct attention to the mutual interpenetration of technology and society. We want to highlight how technology affects social relationships, how social relationships affect technology, and how this changes over time and place. Tool use by humans is found in our earliest histories, but the meanings, symbolism, and power of human tool use has shifted, changed, and multiplied as much as the technologies have. The technosocial highlights this shift and unites technology and society. We live in environments rich in artifacts, ranging from simple tools, clothing, and domestic goods, to the highly complex electronic technologies that dominate our modern information societies. The landscapes of our lives have been changed by human activities. Think about the world we live in: criss-crossed by visible and invisible communication networks, urban skylines in all parts of the globe, and an atmosphere filled with pollutants and air traffic.

We will be working through, both directly and indirectly, several propositions for science and technology studies. We will argue that science and technology can and should be understood as social institutions. This understanding helps to explain the durability of scientific and technical products as well as the social processes that produce them. Secondly, we want to point out that science and technology affect and are affected by the distribution of resources and power in and across societies. Thirdly, we need the tools of historical and anthropological inquiry to understand science and technology effectively. It is important to use the comparisons that other cultures provide as well as the divergences and contradictions within our own cultures to highlight the contingent nature of knowledge.

We have a fundamentally democratic goal: to undermine the ideology of technological determinism, the idea that technological change is inevitable and always "progress," and to make the social institutions of technoscience more responsive to public interests. Our goal is not to be anti-science or anti-technology, but to be careful about accepting and designing technologies and thoughtful in prioritizing scientific research. Critical understanding entails using different patterns of thinking and different technologies, ones that are appropriate to ways of life sustainable on environmental, social, and personal levels. This approach provides tools and frameworks that can articulate what these "sustainable ways of life" might be and whom they are for. Based on these propositions, we will focus this book on questions about whether or not technoscience can be used to improve the quality of things produced, to improve work experiences, to improve the overall quality of life, and what these "improvements" mean and for whom. We do think that it is possible to alter our technoscientific world and we yet do not accept as given that all technoscientific changes are always and necessarily improvements.

While we are not anti-science/technology, we are quite skeptical about the current social relations of the technosciences. We also question whether the overall organization of societies around the world is socially optimal and ecologically sustainable. For science and technology to "work" as institutions that make the world better, they must become socially coherent – that is, allow the same level of skepticism that they apply to the objects they study to be applied to themselves (Wright 1992). Without that level of reflexive scrutiny, scientific claims to knowledge will tend to be self-serving. We are seeking to promote values such as fairness, robustness, and sustainability. We want to open up our technological culture for inspection and ask who controls, who designs, who uses, who benefits, and who loses in relation to the products of technoscientific inquiry, production, distribution, and consumption. This politicizes technoscience, and it will certainly not make the jobs of most scientists and engineers, nor the "average citizen" any easier in the short term. However, in the long run, we see sociologically grounded STS inquiry as contributing to a system of checks and balances in technoscience, in the same way that different branches of the US government maintain checks and balances in politics and law, at least in theory. The result will, we hope, be sciences and technologies that are more reliable (more *true* if you like), because they have withstood a full spectrum of technical and social critique and are more responsive to a broad range of human values.

From the Sociology of Science to Science and Technology Studies

The sociology of science that crystallized in the works of Robert K. Merton and Bernard Barber beginning in the 1930s focused on the social system of science. That is, Merton, Barber, their students, and those they influenced analyzed the reward system of science, the norms of science, social stratification in science, and in general the way the "scientific community" was organized, and how it functioned. We should recognize that Merton and Barber were concerned about demonstrating that science was good for democracy, and democracy good for science in the wake of the excesses and distortions of science under Nazism and Stalinism. Their analysis did not include questions about the "content" of science, that is, about "scientific facts," "truths," and "knowledge" precisely because of the ways the particular historical moment in which they lived and worked shaped the questions they asked.

Beginning in the late 1960s, in the wake of the civil rights, anti-war, feminist, and ecological movements, a new kind of sociology of science began to emerge. The new sociologists were familiar with the criticisms of science and technology developed by "the children of the '60s," and indeed many of them had been activists themselves. At the same time, many of the founders of the field now variously referred to as "science studies," "technology studies," "science and technology studies," and "social studies of science" were themselves originally trained and educated in the physical and natural sciences. Here began the idea that scientific knowledge and technology itself could be studied sociologically. Ethnographic studies of scientific practice and programmatic claims that made mathematics a matter for sociological investigations emerged during the 1970s. By the 1980s, the central dogma of science and technology studies, that science and technology were social constructs, was no longer surprising for those in the field.

However, in the larger realm of society the tension caused by simultaneously (and to varying degrees) treating "science" as the paradigmatic mode of inquiry and as social relations generated an ambivalence about science that has lasted into our own generation (Croissant and Restivo 1995, 67–86). This ambivalence about science has protected it from criticism by critics, theorists and worshippers as they desire to preserve the qualities of inquiry associated with science. Thus, for example, you can find defenders of the purity and truth of mathematics, physics, biology and other sciences in a variety of places. In the professional realm, scientists and philosophers continue to defend traditional Platonist views of mathematics in direct contrast to the efforts of some scientists, mathematicians, and science studies scholars to bring a new critical and theoretical discussion to mathematics (Restivo 1992). For example, one of us traveling in an airplane began a conversation with our neighbor, a cardiologist. When the cardiologist discovers that her follow passenger is a professor of science and technology studies, she cannot stop asking about the "*truth*" of science. The cardiologist's day-to-day reality and her identity is grounded in the *truth* of science and the notion of science as a social phenomenon is a direct challenge to what she does every day and who she is. The social world of the hospital, of organized medical research, sustains the truths produced there, contingent upon people's continued belief and practices in support of those truths. For this cardiologist, recognizing the social dimensions of technoscience created an ambivalence about science that prevented her from engaging in conversation beyond the idea that science is the foundation of truth that the whole world rests upon.

The idea that science is a social construction and a social process is clearly abroad in our intellectual milieu. However, a critical eye needs to be applied to how the idea is being used. For example, Steven Weinberg (1992, 188), a noted physicist, concedes that science is a social process, while staunchly objecting to the idea that scientific products are also social. We do not, as Weinberg supposes, observe that science is a social process and then logically conclude that scientific theories are social. Rather, we demonstrate this sociologically. Weinberg illustrates the absurdity of our claims as sociologists of science with an apparently devastating example. His example is that the right path to a mountain peak is known to exist because it leads the mountain climbers to the peak, not because of the social factors of the expedition. In fact, what Weinberg is ignoring is who made the path, why they made the path, when they made the path, what tools were needed on the path, how the expedition came to be on that path, what power relations were at work to create the organization needed to make a viable path, and the negotiation of the value judgment that one path is more right than others, and that one mountain is more worth climbing than others, and even in the determination of where the "top" is. Or, in other words, he commits the Columbian (Christopher Columbus) fallacy that assumes that things, events, or processes are, or can be, simply, immediately, transparently, phenomenally perhaps, and most importantly a-socially discovered.[7] The sociologically inclined philosopher of science Clifford Hooker recommended some two decades ago that to think about science – or in more general terms, knowledge, inquiry, thinking – we must be at least prepared to criticize (1) particular facts, (2) specific theories, (3) types of theories, (4) conceptual frameworks and perspectives, and (5) the institutions of research and criticism. Every theory of inquiry should include a theory of the intellectual milieu, a theory of critical culture (Hooker 1975, 102–103).

Sociology in general is useful for understanding and explaining aspects of our everyday lives that we do not readily recognize to be social in any way. Some examples of this are love, suicide, religious faith, and science. Within science, mathematics gives the appearance of being the most recalcitrant in the face of the sociologist's toolkit. Historically, mathematics has been the arbiter of the limits of the sociology of knowledge. Restivo (1992) has studied this "hard case" in some depth, sometimes in collaboration with Randall Collins (Collins and Restivo 1983). Some of the confusion about what makes mathematics *the* hard case for sociology is found in general among critics of the sociology of science. For example, explaining the content of mathematics is not a matter of constructing a simple casual link between a mathematical object such as a theorem and a social structure. The sociological method is to look to both "external" contexts and "internal" networks. One common error is to imagine that only "external" milieux hold social influences.

Second, the sociological task is to unpack the social histories and social worlds embodied in objects, such as theorems, proofs, and equations. As Star, Bowker and Newman (n.d.) note:

"Social world" is a term in sociology first coined by Anselm Strauss (1978). It refers to a group of people joined by conventions, language, practices and technologies. It may or may not be contained in a single spatial territory; in the modern world, it typically is not. It is cognate with the notion of community of practice (Lave and Wenger 1991) and with reference groups. It was coined for social analysis in order to speak to strong ties which are neither family nor formal organization, nor voluntary association, and which may be highly geographically dispersed. Examples of social worlds are stamp collectors, rock climbers, activity theorists, and socialist feminists.

Mathematical objects must be treated as things that are produced by or manufactured by social beings through social means in social settings and given social meanings. Mathematics happens in social worlds. There is no reason why an object such as a theorem should be treated any differently than a sculpture, a teapot, or a skyscraper. Only alienated and alienating social worlds could give rise to the idea that mathematical objects are independent, free-standing creations, and that the essence of mathematics is realized purely and only in technical talk. Notations and symbols are tools, materials, and, in general, resources that are socially constructed around social interests and oriented to social goals. They take their meaning from the history of their construction and usage, the ways they are used in the present, the consequences of their usage in and outside of mathematics, and the network of ideas they are part of.

Mathematics, science, and knowledge in general are crucial resources in all societies. Systems of knowledge therefore generally develop and change in ways that serve the interests of the most powerful groups in society. Once societies become stratified, the nature and transmission of knowledge begins to reflect social inequalities. Once knowledge professions emerge, professional boundaries tend to shield practitioners from the realities of their broader social roles even while they define a realm of systematically (institutionally) autonomous work. Science and math curricula are certainly influenced by professional interests and goals, but they are also conditioned by the social functions of educational systems in stratified societies. Latin was the language of schools in Europe for many centuries. By teaching, reading, and writing in Latin the intellectual community controlled who had access to particular knowledge (Wertheim 1997).

Science has provoked some of our most profoundly learned contemporary and ancestral colleagues to describe it as a Machine (C. Wright Mills 1959), a danger to democracy (Paul Feyerabend 1978), and a danger more generally because of its constant desire for certainty. Fredrich Nietzsche (1974, 335) describes science as having a potential for divesting existence of its "rich ambiguity" and reducing life to "a mere exercise for a calculator and an indoor diversion for mathematicians." Nietzsche (1882/1974, 335) goes on to describe science as an idiotic crudity, a mental illness, "the most stupid of all possible interpretations of the world," interpretations of the most superficial aspects of existence, the most apparent things that permit "counting, weighing, seeing, and touching, and nothing more." Now we admit that in the end we might not want to echo these sentiments in just this way, but in the context of understanding the institution of science, all of these labels seem reasonable.

The reductionistic language that Nietzsche points out is also one of the entry points to feminist criticisms of science. Language, culture, and values, are essential, not incidental, to the development of scientific theories (Longino 1990). Science has never been immune to using metaphor, and to extrapolating between contemporary gender orders and scientific understanding. For example, it is clear that classification systems (Schiebinger 1993), or studies of women's capacity to work while menstruating (Martin 2001) have echoed social concerns and stereotypes already present in the larger culture. Current sociobiology cannot seem to get beyond selective analogies from animal models to explain a narrow range of human behaviors (white, middle-class, western families) assumed to be natural and universal (Bleier 2001). Science is invoked to *produce* differences, particularly to demonstrate the "natural" inferiority of women and non-white minorities (Birke 2001; Bleier 2001; Tavris 1992). Obviously, feminists and anti-racist scholars and activists are quite concerned with the deleterious effects of science on women and minorities. And yet, as many have noted, reasoning, critique, and argumentation provide important tools for liberation, and technoscience can aid in identifying and solving problems. The distinction between "the rational" and "the social" is unfounded, and we need not abandon rationality or inquiry, but recognize its partiality (Longino 2002).

As an alternative to "science," we have "thinking," the activity of trying to find something out. Following Nietzsche, in this activity, successes and failures are above all "answers." Our inquiries here are guided by the following sorts of queries: "What did I really experience? What happened in me and around me at that time? Was my reason bright enough? Was my will opposed to all deceptions of the senses and bold in resisting the fantastic?" No more convictions, no more excision of passion and even love. "Objectivity" cannot mean "disinterested contemplation," a "rank absurdity." Let's look for that immense capacity of thinking (and Nietzsche would not be averse to using the term "science" here) "for making new galaxies of joy flare up."

Scientists "are becoming the new villains of Western society" (Overbye 1993). In the 1970s, in a CIBA Foundation symposium, Hubert Bloch (1972, 1), a physician and chair of the symposium, wrote: "And in the minds of many, science . . . has become a most dangerous evil." Earlier still, Goethe, Schiller, and William Blake were "hostile" to Newtonian science. Jonathan Swift scorned the Royal Society. Michel de Montaigne complained about the hubris attending theories of nature. And didn't Montaigne in the sixteenth century echo Socrates' criticisms of the pre-Socratics and even of his pupil Plato? Stephen Toulmin (1972, 24), in fact, suggested that:

> Throughout the last half-millennium, at least, anti-scientific attitudes seem to have peaked at intervals of 130 years or so, if not every 65 or 30–35 years.

It would be a grave error to label these intellectuals "enemies of science" or "anti-science." They all had a commitment to those qualities of thinking and inquiry many educated elites tend to associate exclusively with science, but the institution of science does not hold a monopoly on traits such as rigor, complexity, depth of understanding, and passion for knowledge.

Conclusion

Science worlds are social worlds, and we must ask what kinds of social worlds they are. How do they fit into the larger cultural scheme of things? Whose interests do they serve? What kinds of human beings inhabit science worlds? What sorts of values do science worlds create and sustain? How do science worlds change, how have they changed in the past, and how are they changing today? If we conceive science as some independent free-floating set of methods, theories, and facts – instead of as a social world, or an institution – we might fall into the trap of trying to adopt conventional scientific tools and ways of thinking and working to help solve social, personal, and environmental problems. It is unreasonable to suppose that social reformers and revolutionaries could eliminate *science* from society, and equally unreasonable to suppose that they could force science as we know it today into some "alternative" shape independently of broader social and cultural changes.

It has taken centuries for scholars to recognize the folly of trying to establish absolutely certain grounds for our knowledge and belief systems. However, the effort is often still made. By looking at power relations we can begin to understand why when a social institution offers "certainty" what it is really offering is stability, moral order and the authority to enforce trust – all of these produced with great effort and at great cost. This can be both a positive asset and a negative constraint for a community as determined by the values of the society and the relations between leaders and ordinary people. Transcendentalism (e.g., Platonism), privileged assumptions (e.g., apriorism, and foundationalism) and God are dead. But the protective, awe-inspired, worshipful orientation to science survives. It is this God-inspired foundation upon which western science and religion have been constructed and sustained. When we leave Plato behind, when we finally give up transcendence and foundationalism, we will find ourselves confronted with the end of a certain way of doing inquiry, and with the end of a certain way of living.

The crisis of religious faith that swept across nineteenth century Europe and America has its parallels in the realms of science and logic. Many prominent philosophers, from Rousseau to Nietzsche and writers, from Dostoevsky to Kafka, have questioned our uncritical faith in *reason*. Kafka's assertion in *The Trial*, "Logic is doubtless unshakable, but it cannot withstand a [man] who wants to go on living" would find ready endorsement from Dostoevsky, Nietzsche, and others. These thinkers questioned science, logic, and reason not because they were "relativists" or failed to appreciate the value of inquiry but rather because they appreciated the complexities of social structures and cultures. They were critics of the "Cult of Science" and that cult's intense "faith in science." We advocate ways of talking and thinking about science that do not fix it in the grammar of the ever-present tense. The "Science Is . . ." mantra or chorus fails to capture the social dynamics of science and society – remember the cardiologist's inability to think of science as social. In the case of technology the problem is not so much that it is discussed in the grammar of the ever-present tense but rather in a grammar of inevitable progress.

When we talk about science, truth, logic, technology, and related ideas, we are always talking about social relations. This way of seeing sensitizes us to the progressive and regressive aspects and potentials of words, concepts, and ideas that as social relations can embody inequalities, destroy environments, inhibit individual growth and development and undermine inquiry, as well as solve problems and provide joy.

Notes

1 Individuals, such as Friedrich Nietzsche, Emile Durkheim, Karl Marx, Max Weber, Max Horheimer, Theodor W. Adorno, and Herbert Marcuse were already struggling with the progressive and destructive aspects of science and technology, and their intent was to understand these phenomena in social, cultural, and historical terms. In the early decades of the twentieth century,

Splenger, Mead, Gumplowicz, Wittgenstein, and Fleck helped sustain and develop these ideas in their most radical forms.

2 Colleagues might recognize the threads of symbolic interactionism and pragmatism woven here. See Clarke and Gerson in Becker and McCall (1990) for a review and Clarke and Fujimura (1992) for examples. See also Star (1995b) for a collection informed by the symbolic interactionist and pragmatist convergences.

3 CE means "Common Era," to avoid the ethnocentrism of AD or "Anno Domini" which reflects a specifically European Christian ordering of the world.

4 For example, "science both exemplifies and expands gender stratification within society" (Fox 1999, 441). In this case, science is seen as a masculine enterprise, where values such as independence and control are taken as primary. In addition, women generally are not present at higher levels of scientific hierarchies, despite increasing equality of presence and productivity upon entry into scientific careers. Finally, "science" is used to explain and justify these inequalities, restating the status quo in a self-reinforcing system.

5 See, for discussions and other examples of purifications, Hoffman 1988; Knorr-Cetina 1981; Star 1995a; Latour and Woolgar 1979; Clarke and Fujimura 1992.

6 See also Bowker and Star (1999) for a discussion of the in / visibility of work in nursing, Star and Strauss (1999) for a general discussion, and Shapin (1989) for the classic article on invisible technicians.

7 The "Christopher Columbus" issue brings up a number of interesting topics we will discuss more fully in later chapters. See Turnbull (2000) for a cross-cultural examination of knowledge production, and Watson-Verran and Turnbull (1995) for a review of similar issues, and an articulation of how science is *also* an indigenous knowledge system. Like Columbus "discovering" America, to say that Champlain "discovered" the lake named after him is to ignore the knowledge that the native Iroquois and Algonquin had of their environment. His "discovery" was only relative to the Europeans exploring the northeast of the North American continent. In Australia, the myth of "terra nullius," that the continent was uninhabited by people, and overlooking the many thousands of Aboriginals in various groups living there, was used to rationalize British encroachment on the territory.

Further Reading

Clarke, A. E. and Olesen, V. L. (eds.) (1999). *Revisioning Women, Health, and Healing: Feminist, Cultural, and Technoscience Perspectives*. New York and London, Routledge.

Hughes, T. P. (2004). *Human-Built World: How to Think about Technology and Culture*. Chicago, University of Chicago Press.

Lederman, M. and Bartsch, I. (eds.) (2001). *The Gender and Science Reader*. New York and London, Routledge.

Reid, R. and Traweek, S. (2000). *Doing Science & Culture*. New York, Routledge.

2

Cultures of Science

What is a fact? Once a fact, always a fact? If not, what was a fact before it was a fact? Do facts die? Are scientific facts the same as normal facts? A conventional definition of facts is: those ideas and information that no one questions, that end a conversation or argument and that help establish who is the authority. It is a fact that the earth revolves around the sun – the ancient Greeks knew this. However, by the 1500s European society had "advanced" and now *knew* that the sun revolved around the earth; Galileo and Copernicus came along and shifted the perspective. Now, every school child learns that the earth revolves around the sun. This "fact" stands as a symbol of the ability of science to give us true facts. Isn't it strange how a story like this reinforces our belief that there are true facts and false facts. Why doesn't this story challenge us to think about what a fact is? Why did people believe that the sun revolved around the earth? Had this always been a fact until science proved that it was false? Clearly the message here is that many people understand science's job is to describe the world factually and accurately. How science establishes facts and their accuracy is an important topic of analyses within the discipline of STS.

Before you get dizzy, if you aren't already, let's back up and think critically about the idea of "a fact." A fact is an idea or concept that everyone (or some subset of everyone such as a community or network) accepts as true. It is a fact that she is one meter and eighty centimeters tall. It is a fact that many countries use meters to measure length. What happens to this fact when she is in a country that does not use meters? The idea of germs has been present in North American and European thinking for a long time. Yet the germ theory of disease is a relatively new idea. Does that mean germs did not exist earlier, or don't exist where people have never heard of germs? Science tells us they exist whether or not people know it, that what science uncovers is universal. Is it? Who is uncovering what? We know scientific theories can change but do facts change as well?

Facts are accepted by a community as true in a rather circular way – those who do not accept important facts are excluded from that community. Facts are stable, but their acceptance is established over time, not all-at-once. When first introduced, they may be questioned, but with the passage of time they generally become unquestioned. Facts are practiced and shared by a community of people. You might announce that you are the Queen of Denmark, but it does not become a fact until everyone around you treats and accepts you as the Queen of Denmark. If your immediate circle of friends and acquaintances come to accept you as the queen and no one else does, are you the Queen of Denmark? Virtual reality has clearly shown that one can be a King or Queen on line while being a teenager in a middle-class family in "real reality." If you marry the King of Denmark, then the largest number of people possible, the relevant members of the global community, would accept you as the Queen of Denmark and you would unquestionably be the Queen. But if there is dissent over the legitimacy of the monarchy, are you still the Queen?

A fact has to be named by the community, accepted by the community and practiced by the community. There is an outward spiral of naming, accepting, and practicing facts that moves from persons to an immediate work group and on to networks, communities, societies or nations, regions, ecumenes, and the globe. Note that this is not the source of the fact. Facts originate at the level of social networks and get expressed through persons. At each level of naming, accepting, and practicing, there are criteria that establish and circumscribe those competent to authorize facticity. It might be citizens who are male and landowners or any individuals born on the soil of a given country; it might be professionals credentialed in a certain way. As soon as a fact is no longer accepted and practiced, it loses its status as a fact and is either replaced with a new fact or is laid to rest in the fact dump of history.

Is Pluto a planet? Many of you have probably learned some clever mnemonic like "My Very Educated Mother Just Served Us Nine Pizzas" to help you remember the names and order of the planets from Mercury (closest to the sun) to Venus, Earth, Mars, Jupiter, Saturn, Uranus, Neptune and Pluto (furthest from the sun). In the late 1990s, the International Astronomical Union (IAU), in charge of naming celestial objects, was reported to be collecting votes to demote Pluto from planetary status to "minor planet" or part of the Kuyper belt and its array of trans-Neptunian asteroids. In many respects, Pluto is more like a comet given its icy character and irregular orbit, and several planets have moons larger than Pluto. Its own moon, Charon, is nearly the same size, and reflections from it produced variations in estimates of Pluto's size for a long time. Removing Pluto from the list of planets may on one hand seem to be a trivial exercise in nomenclature. On the other hand, it does reflect finer distinctions and refinements in astronomical theory, particularly of planetary and solar system formation. Some astronomers were concerned that a divergence between popular and expert definitions might weaken the popularity of astronomy, and a decline of public support could translate into a decline in research funding. It would have entailed revision of textbooks, websites, teaching materials, and popular conceptions of the solar system. The debate was not so much resolved as evaded by the IAU in 1999 when it declared there was no movement to demote the planet, despite the great ambiguities in its classification. This case, however, is emblematic of the cultural work that goes into making facts, through establishing consensus and coherence with existing frameworks, and the cultural work that must be done to sustain, revise or discard facts.[1]

Society and Culture

As we begin to introduce the idea that facts have histories and that they are embedded in practices, we begin to demonstrate that some understanding of society and culture is required to fully explain facts. Cultures are constituted of people who hold each other mutually accountable according to the terms of a moral order. One of the major tools of science is the scientific method. Learning the scientific method of observation, hypothesis, experimentation, analysis, and theoretical synthesis creates a common methodology for scientists both expert and lay. The more expert you become, the more you realize that there is no such thing as *the* scientific method. Nonetheless, the myth of a common methodology for doing science provides a common groundwork and a culture that make scientific discourse possible.

Students and even some professionals are often confused or uncertain about the difference between culture and society. Culture is the totality of a group's ways of living, knowing, and believing. Society refers specifically to the ways in which a cultural group organizes and locates itself, and includes the cultural explanations for that social order. Traditionally, cultures have occupied geographically bounded territories. Social institutions within those territories have been the locus of those practices that provide basic goods and services and the resources and activities necessary for reproduction (self-perpetuation). As human populations have grown and spread across large regions of the world and around the world, cultures have emerged that are not geographically bounded. These systems have been variously described as third-cultures, lateralizations, and super-cultures (Restivo 1991, 181–182). In the modern world, banking and the system of air transportation are examples of such cultural systems. Some third-cultures are created out of offensive and defensive efforts by nation-states that are engaged in cold or hot wars with each other. This creates an interesting situation. An international spy may have more in common with his or her enemy counterpart than with the people in his or her home government. In general, third-cultural systems do not have geographical boundaries but they do develop other forms of boundaries – professional boundaries, for example. While they are subsets of one society, they also cut across multiple societies. Science can be thought of as a third culture with its scientific method(s), common language, professional societies, methods of training new scientists, peer review and scientific equipment. In any case, one of the consequences of establishing a societal or cultural boundary is the potential for prejudice or more generally a "we–other" distinction.

Human history is filled with myths, stories, and histories documenting prejudices of various kinds. These prejudices have been reinforced, maintained, and reproduced through cultural beliefs, political authority, religious doctrines, and more recently science. For example, craniometry, the measurement of skull dimension and shape, was very popular in the late eighteenth century and early nineteenth century as a way to rank the differences between presumed racial groups and the sexes in the United States and Europe. The social conventions of this time held that white males were superior to other races and to white women in matters of state, citizenship, politics, knowledge, and authority. Inevitably, the outcomes of crainometry supported the social beliefs that white men have larger brains than other humans (Kaplan and Rogers 2001). The results of the research were then taken by the population at large to be an independent verification of the social beliefs. So were scientists defining or discovering?

22

Worldviews

Worldviews are ways of seeing and interpreting the world from a cultural perspective that provide the tools to categorize and classify the world. Hence if your worldview is based on using technology, you may categorize a new digital technology for viewing movies based upon clarity of the picture, price, availability, and so on. However, if you are the designer of the technology, your categories may overlap with the user and will include additional categories overlooked by the "average" user, such as speed of transmission, resolution, cost of manufacturing, and development time. When we categorize objects, people, or ideas we generally do so using a binary system: good/bad, male/female, hard/soft, technical/non-technical and so on. Two elementary divisions that all cultures use are male/female and young/old. These are categories that originate in our experiences as human beings and our interactions with others. When we exaggerate the difference between hard and soft, in and out, young and old, male and female, we create a semblance of order. As these labels are used repeatedly by a community, they create a common culture, common experiences and common interpretations of events, ideas, and people – worldviews. Once the labels are well established, they reflect the worldview and system of organizing used by the community. When something is out of place in the worldview, it is called dirt, signaling that a boundary or classification has been transgressed. Words like dirt and pollution always point to boundaries.

Social theorists pay attention to worldviews, classification, boundaries and dirt because they provide insights into how the social world works. By studying magic and pseudo-science we can learn about science. By studying people who embrace or reject technology, we can learn about technology's role in the community. We can highlight, analyze, contextualize, understand, and alter our relationships to science and technology by situating them historically, culturally, and socially.

The Social Construction Conjecture

Broadly understood, science and rationality are human activities found in all cultures, in all times and places. More narrowly and in ideological ways, science has been widely viewed (especially by western philosophers and scientists themselves) as defining the rational and objective since the 1700s. On this view, science is considered to be independent of social, economical, political, and subjective influences. Since the late 1960s, however, researchers in science and technology studies have been demonstrating the social, economic, and political influences on science and more significantly on scientific knowledge itself. Social construction, which has become a foil for critics inside and outside of the sociology of science, is little more than basic sociological thinking. When applied to scientific knowledge, however, it challenges rationalist and realist accounts of science that claim logic and evidence (understood to be to some extent transparent ideas which do not require deep reflection and analysis themselves as social constructions) are the primary determinants of validity and theory choice in science.

Scholars in social studies science have emphasized different features of the social construction conjecture depending on their particular interests in theory and research. Some of them stress what a scientist does in making science; others focus more broadly on individual scientists, their work settings (e.g., laboratories or research centers), and social and cultural contexts. One of the consequences of this approach is that it makes us more aware of the relationships between scientific knowledge and centers of power. Taken together, they all serve to draw our attention to social processes and contexts in science, processes and contexts in which scientists organize and give meaning to their observations.

A humanistic trend acknowledges the importance of real human beings in the making of science. The human actor is emotional, experiences conflicts, expresses inconsistencies, and sits as a mediator between science and the wider socio-cultural and political economic contexts of scientific practice and scientific institutions – for example, between biomedicine research and the biomedical technology industries. Relativistic trends arise from recognizing, for example, that when we observe science as a historical unfolding, scientific theories and even scientific facts or truths appear to be relative to specific historical and cultural contexts. A third trend, sometimes referred to as *rhetorical pathos*, is the growing awareness (especially acute in the earliest stages of science studies' development) of problems inherent in the *language* of both science and science studies.

The use of social constructivism to scrutinize modern science amplifies the moment-to-moment, day-to-day, and night-to-night activities and social minutiae (the little things of our everyday lives that often go unnoticed, that we take for granted) of scientists as they go about producing and reproducing scientific culture. This is the significance of the social construction conjecture, and not its alleged relativistic implications. Relativism can be used to affirm or critique social construction. It can be opposed to realism, and this has been one of the flashpoints in the conflicts between scientific realists and social constructionists. A realist believes that there is an independent arbitrator called Nature that exists outside of humans, that facts are distinct from human thought and practice. A relativist believes that the situation or representations cannot be "sorted out" without an outside arbitrator, but there are no universal arbitrators not themselves grounded in a specific historical and intellectual position. In trying to sort out some of the difficulties that emerged in the early days of science studies, Bruno Latour (1987) finessed the problems by claiming that we are as much realists as the scientists whom we study, at least when we make a knowledge claim, and just as relativistic as they are in moments of controversy.

Latour's strategy is only one of the ways science studies researchers have tried to avoid the traps of realism and relativism. Other strategies include emphasizing negotiated meanings as opposed to causal analyses, and arguing that relativism is simply an acknowledgment that we haven't gotten to the truth of the matter yet. Many of these strategies end up undermining the powerful causal and critically realist assumptions that ground the sociological imagination.

One of the ways to avoid the relativism–realism dilemma is to focus on the meanings and practices of scientists doing science. This is not necessarily a guaranteed way to avoid the relativity problem because there are some philosophers and scientists who are inclined to interpret any social approach to science as inherently relativistic. The most helpful move to make in this situation, then, might be to recall that relativism was not meant to oppose realism but rather to oppose absolutism. Somehow, critics and colleagues too easily forget this. They also ignore or overlook the fact that two of the founders of the "relativistic" sociology of science, Barry Barnes and David Bloor, defined relativism as "disinterested inquiry," a classical definition of science.

Social construction has been viewed by some science studies scholars as a matter of applying the known and successful methods and theories of the sciences to science itself. It can, however, be something more – a multi-purpose tool that allows for the possibility of asking different questions and observing differently, one that can be used by different people with different backgrounds, cultures and socio-political positions whose voices and views on the nature of science might have been silenced or ignored in the past. It may still be middle to upper-class, white, well-educated people doing the research but the voice of a social worker may appear on the same page as a medical doctor without deference to the doctor because he is a white male and the social worker a black female (for example). The works of Emily Martin (1994), David Hess (1999), and

Rayna Rapp (1999) are examples of the emergence of an "equity of voices" movement in science studies and scientific research.

Social constructivism allows for a change in perspective. When the status of science is that of a privileged form of inquiry, only one sort of question can be asked: "what" questions. When science is seen as a discourse, a different sort of question can be asked.

> If science is a discourse whose status as privileged inquiry within the social formation is historically rather than naturally constituted, its autonomy is always mediated and therefore relative to its position within the social formation of which it is a part. Its place is constantly renegotiated with other power centers, and the degree of its "freedom" is always understood in context. (Aronowitz 1988, 300)

Aronowitz's perspective points us to questions about science in relation to the other power centers, such as organized religion, the nation-state, and economic organizations. These are questions about implicit and explicit power relations, questions about inclusion and exclusion, questions about context and contextual change. Herein lies the power of social construction as a creation of the sociological imagination.

Feminism and Science Studies

Feminism is contributing to science studies by providing clear examples of social constructivism, by demonstrating the use of power, domination and language in science, and by creating and applying new methodologies for the study of science. The predominant theoretical framework feminists engage to study or critique science is social construction and related inquiry into the social context of science (Rose 1994). By studying networks of actors, their practices and the construction of scientific facts, feminists are demonstrating sexist and racist bias in science and exploring the relationship between culture, difference, and science.

Feminism gained momentum and strength in the mid-twentieth century United States with the civil rights movement and the advent of the 1950s, 1960s, and 1970s as women gained widespread access to all levels of higher learning. It was not that long ago that it was believed that education would shrink or harm a woman's reproductive organs and that it was thought unnecessary and unladylike for women to seek an education. As more and more women entered the universities, they began to raise the question, and prompt others to ask why there were so few women in the sciences. Questions about equitable participation opened up analyses of issues of objectivity, rationality, purity, and truthfulness in science. Feminism gained strength in a cultural moment fueled by returning Vietnam veterans, civil rights activism, the cold war and the threat of nuclear war, and a disillusionment with science and technology as aids in bettering the human condition.

Feminists have been contributing for decades to our understanding of the sexist and racist biases in science and the gendered cultural contexts of scientific practice. They have debated the relative virtues of focusing on (1) whether sex differences are socially or biologically grounded or on (2) equity and equality issues. Some researchers have demonstrated the masculine bias in how female sexuality is portrayed, others the cultural assumptions about what is "normal" in discussions of hetero- and homosexuality. A masculine bias has been demonstrated in evolutionary studies and endocrinological studies of behavioral sex differences. Research has been carried out on the effects of assumptions about the inferiority of women on theories of human reproduction. Feminist scientists have shown that gender biases lead to gender associations in the study of cells and their components. Feminists were not the first researchers to reject the traditional perspective on the nature of science, but they were the first to carefully study, document, and demonstrate the many ways in which sexism affects and has affected the nature and practice of science.[2]

Using a social constructivist framework, feminists are revealing sexist bias in science's explanation of sexual differences, reproductive theory, and medicine. All of these scientific explanations deal directly with people's understanding of themselves and directly reinforce the sexist bias of the culture. Sexual bias occurs throughout science, but it is most blatant in the human sciences and they have been the starting point for most feminists. Feminists, and other scholars, will continue to research and articulate sexual bias in other science disciplines. It has become increasingly clear that problems of gender are not about women but about culture and the ethos of science. Culturally, for example, we in the west seem to be obsessed with difference. Since the beginning of the "scientific revolution" in the seventeenth century differences between male and female bodies have been viewed by various observers inside and outside of science as natural differences. It is a small step to move from natural differences to natural laws that ground the gendered inequalities that divide men and women in terms of power and privilege. If difference is rooted in natural law, it supports and sustains the status quo. If difference is produced by society, in terms of opportunity and access to all sorts of institutions, then the status quo is not a "natural" state of affairs at all.

Another approach used by feminists to critique science is to address issues of power, domination and politics (Harding 1986; Schiebinger 1993; Tuana 1989; Rose 1994). In the late twentieth century researchers and social theorists have begun to study science systematically to show how the kind of science and thinking we do is influenced and shaped by who does it, who pays for it, and who is asking the questions. In other words, the people in power are the ones who determine the type of knowledge created. Middle to upper-class, well-educated, white males generally practice the science that has determined that women and minorities are incapable of performing science (Harding 1986). This conveniently eliminates other perspectives that might have questioned modern science (Schiebinger 1993). Feminist social theory, as well, is not immune to the risks of universalizing, silencing, and marginalizing (hooks 1982). Feminists are discussing and developing methodologies and theories that try to incorporate and acknowledge the empowerment and disempowerment potentials of discourse.

The changes and responses of society to modern reproductive medicine have been a major focus of STS research since the 1970s. With the introduction of new reproductive technologies, parents and healthcare professionals have had to face new ethical and moral dilemmas. Amniocentesis began to be widely used in the United States in the late 1960s and was first used to search for chromosomal errors. In studying women's choices about amniocentesis during pregnancy, Rayna Rapp (1990, 33) found that when "the hegemonic discourse of science encounters cultural differences of nationality, ethnicity, or religion [it] often chooses to reduce them to the level of individual defensiveness." Modern science has little tolerance for or ability to cope with cultural difference because it has no way to make it fit with the assumption of universalism. Science experts had developed and implemented new procedures to identify certain genetic conditions before birth with the assumption that it was better to know before birth. In the case of amniocentesis, it also assigned the risk of fetal loss or miscarriage caused by the procedure to the expected likelihood of revealing an abnormality. As the technique has improved this has led to lowering the age at which women are pressured to use amniocentesis and to make decisions about carrying a potentially "defective" fetus to term or having an abortion. It has also led to a calculus that equates a probability of death with a probability of a genetic disorder. However, for potential parents it is never simply a matter of just knowing, but of deciding how to evaluate, act upon, understand, and incorporate the information into their lived experience.

Feminists have also raised questions about "neutral" science policy and the not-so-neutral implication for the "other" (Reid 1987). As long as AIDS was perceived as a disease affecting "others" (primarily male homosexuals and IV drug users), it was not taken seriously by the medical community and research agencies (Epstein 1996). That changed dramatically when people who were not "others" were identified as infected. This can be explained by thinking about who is making what decisions with what particular values. When men, particularly powerful white men, are the unmarked category to which other groups are compared and treated as "other," all other groups are seen as "special interests" while the interests of the dominant group are taken to be neutral. That means that the implications of science and technology policy decisions are taken to be neutral even though they represent the interests of their originating groups, who claim to speak for everyone. Sometimes this can be of benefit to everyone, as when cancer was identified as an important disease to study. However, when the study of prostate cancer is given more attention and funding than uterine cancer, only half the population is benefiting. Perhaps some general knowledge of cancer might come out of the study of prostate cancer, but that might not be of direct benefit to women.

By paying attention to languages, metaphors, cultural artifacts, and symbols, feminist theorists have broadened the typical explorations of science to illustrate its social construction. One of the main goals of science is the production of knowledge and information. This production process is not possible without a language that communicates meanings. We think, talk, and do, using language (Lakoff and Johnson 1980; Rorty 1987). "Doing" is influenced by how we conceptualize the world and it is through language and metaphor that we describe, understand and interpret experience. So one of the major contributions of feminist social theories has been to pay attention to the language and metaphors of science. Feminists have highlighted how the language of science (like all languages) is gendered, racialized, and created by users and producers within a network. Thus the non-neutrality of science is illustrated by paying attention to what it says and does not say, by what it does and does not research, and by what it does and does not discover. In re-evaluating, noticing, and articulating the rules governing the acceptability of language and discourse for gendered, raced, and over-looked content, feminists have broadened the questions being asked, researched, and funded to include multiple interests (Irigaray 1989). Aids research, health research, and reproductive medicine have clearly been enriched by applying feminist perspectives (Stephan 1993; Epstein 1996; Clarke 1998; Blizzard 2005).

> Women are trapped in an androcentric world . . . one in which language and meaning have been constructed around androcentric goals and enterprises. We've had troubles with language all along. (Ginzberg 1989, 81)

In order to deal with the troubles of our current language, Evelyn Fox Keller (1989, 44) suggests that a new language that acknowledges difference would be more advantageous for knowledge production:

> We need a language that enables us to conceptually and perceptually negotiate our way between sameness and opposition, that permits the recognition of kinship in difference and of difference among kin: a language that encodes respect for difference, particularity, alterity without repudiating the underlying affinity that is the first prerequisite for knowledge.

The jury may be out about the possibility of creating a "new" language or reshaping the present one. However, changes have occurred in the language of the twenty-first century as we have introduced gender neutral pronouns, "politically correct" terms, and developed concerns about inclusion and exclusion in everyday speech patterns. Of course, there is resistance in the form of jokes and exaggerations, but this resistance highlights the boundary transgressions that such new speech patterns mark in relation to traditional social patterns.

Feminism is enriching science studies with its innovative use of methodology. Sharon Traweek (1988, 1) applied anthropological methodology to two communities of high energy physicists and asked "how" questions such as: how have they forged a research community for themselves, how do they turn novices into physicists, and how does their community work to produce knowledge. Under the scrutiny of Sally G. Allen and Joanna Hubbs (1987) the imagery and symbolization of alchemy is interpreted as man/science's mastery of woman/nature. Helen E. Longino (1990) uses a process-based approach to characterize feminist science instead of the traditional content-based approach. Ruth Ginzberg (1989, 69) reconceptualizes science with her analysis of midwifery: "There has been gynocentric science all along, but . . . we often fail to recognize it as gynocentric science because it traditionally has not been awarded the honorific label of 'Science'." Emily Martin's (1994, 7) methodology is in direct opposition to typical social science perspectives that aim to simplify the object or subject being studied:

> I assume as a starting point that seeing science as an active agent in a culture that passively acquiesces does not provide an adequately complex view of how scientific knowledge operates in a social world. I deliberately

cross back and forth across the borders between the institutions in which scientists produce knowledge and the wider society.

Feminists were among the first to raise questions about the cultural constructions of "objectivity" and helped bring this problem into the mainstream of science studies research. Feminists have identified and asked new questions that the mainstream then incorporates. This has been a major role of feminism and most likely will continue. Innovation often occurs in the margins and how its liberating potential will work for science is still open for debate. Some suggest it as a means to clarify and enrich science and to be more objective while others understand it as a way to create a new science or a more inclusive science and even as a new way to understand objectivity. Feminist sciences would be sciences consistent with feminist values and they would acknowledge the interplay of principles. They would also be sciences that incorporated critical self-reflection into the traditional scientific methods of rational and empirical inquiry (Keller 1985) while also critically gazing at gender and the outsider's experience as valid sources of knowledge. Feminists' goals are to create a science that is not based upon domination, that rejects claims for pure objectivity, and that recognizes other logics (Haraway 1989).

Feminism is changing modern science and studies of science internally and externally by women's presence in the sciences, feminist critiques, and feminist theories. There is hope that space is being created for women, other silent people, and their expertise inside and outside of science. Feminism is multifaceted with its awareness of difference, power relations, domination, language, and the need to create new methodologies. The contributions of feminism have only begun to ripple through the mainstream and have yet to achieve their potential for influencing science and science studies. Most likely, once feminism is absorbed into the mainstream, there will be a new group on the margins asking new questions and challenging the current status quo. Some feminists are skeptical about science studies' openness to the theories and criticisms of feminist scholars (e.g., Loughlin 1993). In the same way that the pull of scholasticism has created a gap between academic feminism and social movements on the ground, it is clear that the liberation potential of feminisms in science and science studies can be constrained by institutional barriers.

Gaining knowledge allows us to turn up new ways of looking at things, and new kinds of things at which to look. We are constantly reminded that it is not the task of inquiry to only lay down answers, but rather to open up new paths of discourse, to reveal new ways to deal with situations, and new kinds of connections in the world. Conclusions are always provisional, and various solutions may co-exist without canceling out (Heldke 1989). Studying science as a social construction opens up the pathways to new ways of looking and to understanding how knowledge and difference are constructed, applied, and maintained.

Technology in Motion

While constructing scientific knowledge is a harder case for STS practitioners, the socio-political construction of technology is more obvious, even if we collectively manage to forget it most of the time. Technological diffusion is the process by which a new artifact or process, an invention, moves into use. An innovation is an invention that has become integrated into society. Innovations solve problems and meet needs (whether at the level of basic survival or manufactured ones as society grows and expands) and often have unanticipated consequences. Sometimes inventions are the result of planned innovation, deliberate searches for new products or solutions to problems, and sometimes of serendipity. Many (and maybe even most or nearly all) inventions and scientific discoveries are what sociologists of science call *multiples*; that is, they appear in more than one place at roughly the same time. It is often unclear where inventions come from. While in western societies we like to attribute invention to individual effort and genius, most inventions are the product of simultaneous discoveries and collaborative activities. Even individual efforts are collective in a sense, since individuals are social constructions and carry society and culture everywhere they go. Simultaneous inventions lead to priority disputes, in part because we want simple attributions. For example, the HIV virus was discovered as a result of a long collaboration between researchers in the US and France. As the results of their work became clearer, researchers began to argue about priority.

This argument was about the scientific and intellectual property privileges that would arise from being declared "first" and winners of the race. Similarly, while many have heard of Watson and Crick and give them credit for the "discovery" of the structure of DNA, most people have never heard of Rosalind Franklin, the crystallographer who provided much of the data Watson and Crick relied upon, and was close herself to understanding the structure of that molecule (Maddox 2002). Besides these problems of collaboration and priority, people working on what looks superficially like the "same" problem, and using similar strategies and resources, often have very different ideas about the nature of the problem.

This leads us to the issue of competing alternatives. Often, there are several different solutions to a problem or need that are being explored. At any given time, there is competition among inventions and their supporters, to try to move a product from an idea to a widely distributed innovation. Many times, multiple solutions will co-exist. For example, the automobile competes with bicycles, rail, mass transit, and walking as a form of transportation. Or postal systems, telecommunications, and courier services are all viable options for moving information from place to place. Similarly, while jet engines can move heavy aircraft at high speed very efficiently, their efficiency is limited to longer flights. Thus, small propeller-driven aircraft have a solid niche in the airline industry for short flights and small loads. It is important to understand that there are no "pure" technical criteria that will tell you which alternative might win the race to become an accepted innovation, because this depends on the various contexts in which an innovation might be applied as well as on preferences and power (as we will show below).

One of the most difficult tasks for a potential inventor is to clearly articulate the problem to be solved and the audience to be addressed. An inventor needs to describe the performance characteristics for a technology, to be able to describe exactly what the technology does, and how. For example, in the early history of radio, inventors were trying to figure out how to make radio a replacement for the telegraph, to create a wireless form of point-to-point communication. The characteristic of radio waves to spread out, and the possibility of broadcast were, at first, imagined as a problem to be solved, with inventors filing patents on devices to attempt to make radio like telegraphy. Eventually, however, people got the idea that radio's broadcast ability was an asset, not a problem, and the technological trajectory and research efforts shifted. Soon after, the modern radio industry emerged.

An innovation does not have a single performance characteristic, but instead there are many, based on different potential users' needs. For example, two cars could be compared based on gas mileage, noise, style, power, and reliability, while the automobile itself can be compared with light rail for efficiency with respect to traffic volume, trip duration, and environmental impact. The performance characteristics of a technology partly determine which alternative will "win" and be widely accepted. However, there is nothing to tell us which characteristics are most important: is it style, mileage, color, or power for a car? The relative importance of various characteristics are socially decided and culturally grounded. For example, environmentalists think that pollution is the main criterion by which a car should be chosen (if choosing a car at all), while most teenagers think that power and style matter more.

Closure is the name for the point at which all of the social groups working (or competing) to define a new technology come (or are forced) to an agreement as to what the new *thing* is. It is an analogous process to deciding what a *fact* is. For example, bicycles today look very different from what they looked like at the turn of the twentieth century. The "ordinary" bicycle with its high front wheel met the objectives of "young men of means and nerve" to show off their new technology, their money and leisure time to ride, and their skills and daring (cf. Bijker 1987, 1995). This version of the bicycle was not popular with female riders, or with older or younger riders, and was seen by them as a "non-working" technology. The "safety" bicycle, what we now consider a *real* bicycle, was seen initially as a compromised design. However, the fact that designers today still invent bicycles that more or less conform to our conventional definitions of "bicycleness" is an example of closure, where there is a fairly uniform convergence on a "paradigmatic" or "normal" configuration for a technology. We all *know* what a *real* bicycle is supposed to look like, despite variations in components and style. Consider, however, that recumbent bicycles are biomechanically more efficient and easier on most people's backs and shoulders. Culturally, however, at least in the US, recumbents seem out of place given the prevailing definition of a *real* bicycle.

In establishing closure around an innovation and identifying it as the paradigmatic frame or model, various things can occur. Standardization is one part of this process; the government and related agencies or industry groups agree on specifications and features of a new technology. For example, 120 volts alternating-current power is standardized in North America while in South America you can find 110, 120 and 220 volts as standards. Various processes can lead to standards, although the most obvious one is through monopoly relations, where one producer or vendor determines what the basic performance of a technology will be. Similarly, an oligopoly or cartel, a small group of manufacturers, can agree on standards. Other kinds of power plays can also be found, as in the case of computer operating systems.

As a case study, let us ask: why did the private internal combustion engine "win" as the US's paradigmatic mode of transportation? It is now the mode against which alternatives are compared, rather than the other way around. At the turn of the twentieth century, the electric car was as viable as the gasoline-powered car, and in fact electric cars were setting speed records in competition with gasoline cars. There were still experiments with steam-powered automobiles at the time. Similarly, trolley and other public transportation systems are more thoroughly integrated into other societies than in most US cities, so the question of private versus public transportation is also up for discussion. If you were an investor in 1900, it would have been very difficult to decide whether to put your money into private or public transportation, and to decide on what power system might be the most technologically sound and profitable.

As a consumer, the desirability of a transportation system varied based on your needs. For rural families, the car spoke of freedom from farm life and the possibility of participating in urban culture. For city-dwellers, the automobile opened up the suburbs, and the possibility of travel and vacation. However, public transportation could have done these things too, and at the time the railways were the most established means of moving products around. Culturally, the private automobile fit in well with ideas about individualism and freedom already pervasive in US culture. These values were also being manipulated by the early advertising industry. Touring, simply going out in the countryside (there were not many paved roads let alone an interstate highway system), was a popular activity among elite and middle-class men.

Several social institutions had to line up to make the automobile available for consumers. Mass production, thanks to Henry Ford, helped make cars more affordable. But even then, banks and consumers had to be convinced of the desirability of consumer debt, for cars were still quite expensive for middle-class, working-class, and rural consumers. Until that time, a home or farm mortgage was the only debt that was allowable, in law and custom. Steel and oil producers had to reorganize and did so quickly when they saw the profitability of automobiles. Similarly, consider the feedback loop between the development of suburbs and cars, and their desirability for homeowners. No one would be interested in living in a suburb unless there was reliable transportation to and from urban centers of employment and entertainment, and yet until there were suburbs, there was really no need for most people to own an automobile on a daily basis. Many trolley lines spread to early suburbs in anticipation of growth before the spread of automobile culture.

The electric car was a popular technology. At the turn of the twentieth century, it had distinct advantages in reliability over the gasoline-powered car. There were a number of manufacturers producing cars for individuals and for industrial transport. Most homes did not have electricity at the turn of the century; neither did they have garages for any sort of car. Gasoline stations were few and far between as well. While we have a pervasive myth that the batteries for electric cars were somehow deficient, we have to think carefully about the performance characteristics. For short trips, the electric automobile was cleaner, quieter, smoother, and more reliable to run. You also did not need to crank an electric car to start it. Thus, the need for strength and the risks of broken arms from crank starting the early gasoline cars were undesirable for many users. The electric car had heating and cooling systems, which many gasoline powered cars of the time did not. In many elite families, the electric car was preferred by women drivers for their various activities, while men had gasoline powered cars for touring. For example, Mrs. Henry Ford had a small electric car as did Mrs. Thomas Edison. However, middle-class families could only afford one car (Schiffer et al. 1994).

Given that middle-class women at that point in time were economically dependent on their husbands, men's priorities generally ruled economic decision making, and thus if the family was to have an automobile, it would be a gasoline-powered automobile. Thus, an extensive market for the electric car did not open up, and as a result improvements in the electric car did not occur as quickly as for the gasoline-powered automobile. World War I saw the demise of most of the electric car firms, and the Great Depression finished off the rest of them.

Based on this story, it is not some sort of simple "best" technology that always wins. It is a matter of the cultural ideals, economic relations, and social power relations that define what counts as best, for whom, and for whose preferred activities and values. Even today, when most trips by car are at less than 40 miles per hour and less than 30 minutes in duration, one could argue for the superiority of an electric car or a public transportation system. However, the manufacturing infrastructure is not available for electric cars, and more importantly, cultural beliefs about what a "real" car should be able to do, and about the (un)desirability of public transportation, override the technical superiority of electric cars or public transportation for many contexts.

Pre-scientific: You or Me?

David Bloor (1991) has pointed out that knowledge looks different from different angles. It is like a landscape. "Approach it from an unexpected route, glimpse it from an unusual vantage point, and at first it may not be recognizable" (Bloor 1991, 81–112). Or in other words, if you change your worldview, you might see something differently. However, is it easy to change your worldview consciously; in what ways do our cultural lenses enhance, shape, or clarify what we see?

One of the first big questions for science studies scholars was why did modern science come out of Europe and the Enlightenment. Scholars have documented the presence of science or science-like knowledge in the history of Egypt and China before the seventeenth century. However, it did not grow and expand like modern science did in the eighteenth and nineteenth century in western countries. Sub-saharan Africa was virtually invisible in this conversation and generally assumed to be too uncivilized and primitive to have any sort of science or technology. We now recognize that there was a major iron production system in Africa beginning around the middle of the first millennium BCE. African science and technology was judged inferior to western science and technology by European colonizers in the early 1800s. This judgment reflected a political and economic agenda because it is easier to justify colonizing a people if you perceive them to be inferior and in need of your superior knowledge, particularly when you are in need of their resources (cf. Adas 1989). Thus in the late 1960s when historian Robin Horton (1967) addressed African traditional thought and compared it to western science, he was doing theorizing with a progressive model: science evolves out of traditional thought and improves as it evolves. His goal was to demonstrate that African traditional thought precedes western science in the evolution of theories. This was one of the first efforts to compare western science to an alternative knowledge system and was widely read in the years leading up to the emergence of science studies. A critical examination of Horton's ideas about African knowledge production in connection with a discussion of what theory is with sociologist David Bloor (1991) highlight the ideas of worldviews, social construction of knowledge, and the role of culture in shaping ideas.

The aim of theory "is to grasp significant truths about the world" (Bloor 1991, 51) and to explain and regulate the general principles at work in our world. A theory of knowledge, according to Horton, shows causality, analyzes, synthesizes, and evolves. Western science and traditional African thought both seek to understand, and classify, experience. The tools of one may be atoms and the tools of the other gods, but they both invoke theory to elucidate and organize the world. "Like atoms, molecules and waves, then, the gods serve to introduce unity into diversity, simplicity into complexity, order into disorder, regularity into anomaly" (Horton 1967, 52).

Theory allows the physician and the medicine man to diagnose an illness beyond the constraints of the obvious symptoms. "Both are making," Horton (1967, 54) explains, "the same use of theory to transcend the limited vision of natural causes provided by common sense." Theory is a more abstract and potentially generalizeable form of causal explanation than common sense causality. Theory is preferable when understanding is required beyond the confines of a common sense interpretation. Horton (1967, 60) describes theories as coming from and returning to the "world of things and people." He suggests "that in traditional Africa relations between common sense and theory are essentially the same as they are in Europe." Theories help to explain the unfamiliar and make it familiar. Western culture relies on the familiar as the "impersonal idiom" to explain the unknown, while traditional culture uses the "personal idiom" to understand the unfamiliar. Western science uses its definition of objectivity and distance from individuals to give it authority, while African thought depends upon the speaker and his/her community to provide the authority for the knowledge. Horton describes both systems as picking and choosing what is appropriate to support their theories. Within the culture, he notes that modifications will be made to a theoretical model to enlarge it and manipulate it to fit current circumstances. Bloor is in agreement with Horton that "sometimes scientists will calculate what more is to be gained by conformity to normal procedures and theories than by deviance" (Bloor 1991, 44).

Horton (1967, 68) recognizes "the theoretical models of traditional African thought [as] the products of developmental processes comparable to those affecting the models of the sciences." Horton declares that since African traditional thought shares so many theoretical characteristics with western science, then it must be a precursor to western science. He labels it pre-scientific. He is unable to interpret the similarities as proof that African traditional thought is a "science" or worldview in its own right. His worldview or cultural lenses prevented him from seeing African thought as anything but pre-scientific.

In discussing the Azande view of witchcraft, Bloor states that their logic is not comparable to the western perspective. He states that Azande logic "is quite a different game, that does not have a natural extension into our game" (Bloor 1991, 140). Or, in other words, Azande logic is a different sort of language game than western science. We can take this further by saying that the western game is quite different from that of the Azande and does not naturally extend into their game. Horton's analysis of Azande logic assumes that they are the same language game and thus comparable. His article is an example of western science using its theory to impose and declare its superiority over another in order to eliminate alternative theories. In critiquing Horton's argument and theory, we run the risk of being accused of doing to Horton what he did to traditional African thought. Our intention here is not to declare a general superiority but to

explore avenues overlooked, erased, or hidden, in the hopes of not repli-
cating the mistakes found in the earlier literature. We know much more
today about the cultural and local embeddedness of the sciences (cf. Restivo
1992; Verran 2001).

In positing African thought as pre-scientific, Horton does not recognize
the points of incommensurablility of western science and African traditional thought. It is traditional societies, he explains, not western society
that lack an awareness of alternatives. For Horton, western science's beliefs are not absolutes and are open to some questioning. In traditional
thought, in Horton's view, beliefs are absolutes and unquestioned. Horton
acknowledges that a closed society can begin to open and to recognize
alternatives. When this happens, beliefs become less sacred. When a
culture begins to move away from sacred beliefs, Horton claims that a
transition from traditional thought to scientific thought has begun. For
Bloor (1991), it is not possible to move away from sacred beliefs because
all knowledge has sacred and profane aspects. The sacred aspects of
science are determined by the culture. Horton does not recognize that all
knowledge has both aspects present and that they are determined from
within the culture.

"Science," Bloor (1991, 57) writes, "clearly captures many of the values
which anyone with a commitment to science would naturally want to
endorse." Horton is captivated by science and clearly endorses it. He
demonstrates this throughout his work but makes his point most clearly
when he declares that the only true means of success in human affairs is
through the impersonal idiom of western science. The impersonal model
he advocates makes it impossible for him to recognize the influence of
social structures, cultures, and ideologies on knowledge. The similarities
between African thought and western science that Horton reveals do not
lead him to recognize that different cultural factors are at work. Instead,
he concludes that African thought is "pre-scientific." What are we to make
of Horton's theory of knowledge in the light of Bloor's claim that epis-
temologies are "reflections of social ideologies" or the concept of theories
as worldviews (Restivo 1994). Bloor (1991, 53) notes "knowledge has to
be gathered, organized and sustained, transmitted and distributed." All of
these activities tie knowledge to institutions, power, and authority.

Bloor's (1991) approach to knowledge includes the recognition that
culture provides the framework for individual experience. Thus for him,
individuals in different cultures will have different experiences of the world
with different interpretations. The theoretical component of knowledge
is a social component according to Bloor. Horton does not discuss the
divergent cultures that produced western science and African traditional

thought. He hints at it, but the closest he comes to talking about culture is in differentiating between personal and impersonal idioms. Then he marks the impersonal as superior to the personal in a way that reflects the values of contemporary science and society.

Bloor (1991) describes truth, belief, judgment, or affirmation as efforts to capture reality and to portray how things stand in the world. Bloor uses beliefs as an indicator of the theory working to explain reality. Horton (1967) never addresses the issue of whether or how traditional African thought works to explain reality in traditional African society. He judges traditional thought as inferior to western science without evaluating it in its own environment.

Looking at an African logic, from Yoruba, as neither a precursor to, nor completely incommensurable with western logic, and in its own environment is exactly the task that Helen Verran (2001) has taken up. Her work is convergent with what is becoming known as post-humanism (Barad 2003), an attempt to make the simultaneously material and symbolic actions of humans-in-the-world open to inquiry without referring to foundational categories. Horton's major error, in his asymmetric analysis of African traditional thought, was to take as given or *a priori* the categories of analysis, logic, and certainty, and an essentialist definition of science. A purely relativistic or symmetrical analysis errs by making knowledge systems completely incommensurable, and essentializing each. Verran's objective is to help us see that what appear to be foundational categories, such as number and amount, are the outcomes of collectivities and actions, grounded in culture, ritual and routine, and sustained by political, economic and ecological relations.

If African traditional thought is pre-scientific, why has African modern thought not created its own "western" science? Bloor's (1991: 157) answer would be that "ideas of knowledge are based on social images, that logical necessity is a species of moral obligation, and that objectivity is a-social phenomenon." Thus, western science is a construct of western culture and not a construct of African culture. There is no reason to expect that African culture will or should produce western science because its social institutions, power relations, authority, and knowledge structures are not western. Horton is able to argue that African traditional thought is pre-scientific because he does not address the issues of cultural norms, values, or social structure. Horton would not accept that objectivity is a-social phenomenon. For him, the essence of objectivity is that it is an efficient tool that explains and predicts. Horton is able to indulge his view of what is natural by gazing at a small piece of reality. This small piece is manipulated and adjusted to fit his theory according to his own social ideology.

So, as we manipulate and adjust a small piece of reality into contemporary theory, we find it helpful to remember that "science is our form of knowledge" (Bloor 1991, 160), that "our" is a reflection of social and cultural institutions, and that western science is not the only form of knowledge. The way western science and philosophy have traditionally conceived of mind, for example, tells us more about our culture and systems of knowledge and beliefs *about mind* than it does about what mind "really" is.

Mind and Society

Early on in the history of STS, practitioners recognized the significance of the "hard case." A hard case in this context is a subject matter that doesn't seem on first glance to be amenable to social analysis. Mathematics was traditionally the arbiter of the limits of the sociology of knowledge. It was the ultimate hard case because it was difficult to show how the universality of mathematics could be explained by a-social constructionist approach. The sociology of mathematics changed all this. Today, mind or mentality is probably the hardest hard case. Traditional and prevailing approaches to mind and mentality in general center on the brain. Mentality is viewed as either caused by or identical with brain processes. Given this perspective, John Searle (1984, 18) could argue that "Pains and all other mental phenomena are just features of the brain (and perhaps the rest of the central nervous system)." But Durkheim's analysis of different degrees of social solidarity and the social construction of individuality suggests a culturological conjecture on pain: the extent to which a person feels pain depends in part on the kind of culture s/he is a product of, and in particular the nature and levels of social solidarity in the social groups s/he belongs to. Furthermore, the symbolism of the pain experience in its cultural context is also a determinant of felt pain. Pain has a context of use, a grammar. Such a conjecture was indeed already formulated by Nietzsche (1887/1956, 199–200) in *The Genealogy of Morals*. Wittgenstein's (1953/2001) writings on pain in his *Philosophical Investigations* provide additional ingredients for a social theory of mind based on the role of language in our pain narratives. But Searle, while he invokes the social, does not know how to mobilize it theoretically, and so argues that consciousness is *caused* by brain processes. We will see as we proceed why this claim that has seemed so reasonable for so long must be reconsidered in light of what we know about the relationship between social life and consciousness, and what we are learning about social life and the brain.

Cognitive psychologists tend to view the mind as a set of mental representations. These representations are then posited to be causes behind an individual's ability to "plan, remember and respond flexibly to the environment" (Byrne 1991, 46). Cognitivists also have a tendency to equate cognition and consciousness. But Nietzsche long ago had the insight that consciousness is a social phenomenon. He was one of a number of classical social theorists who had pioneering insights into the social nature of mentality.

We can approach the history of discourse on mind in terms of conflicts and dialogues between rationalists and empiricists, behaviorists and empiricists, ethologists and behaviorists. This is a history punctuated by many famous names in the history of philosophy from Descartes to Kant. Many of their ideas continue to find representatives among contemporary theorists of mind (Fodor 1983, for example, carries Kant into our own time).[3]

Why is it we "locate" mind, thinking, and consciousness inside heads? Certainly in the west, mentalities and the emotions have been associated with the brain and the heart since at least the time of the ancient Greeks. More recently, localizationalist physicians and neuroscientists have reinforced the idea that mentalities are "in the head" (Star, 1989). On the other hand, in sociological perspective, mentalities are not produced out of or in states of consciousness; they are not products, certainly not simple products, of the evolution of the brain and brain states. Rather, they are by-products or correlates of social interactions and social situations. This implies that the "unconscious" and the "subconscious" are misnomers for the generative power of social life for our mentalities – and our emotions. The unconscious is like God – a set of cultural mechanisms for translation and transference that objectify experiences that exist as the products of social interactions, *not* their causes. The thesis here is that social activities are translated into primitive thought "acts," and must meet some filter test to pass through into our awareness That is, our upbringing (socialization) puts filters, so to speak, in place that help to control the thoughts and behaviors we experience and exhibit (cf. Wertsch 1991, 26–27 and see Vygotsky 1978, 1986; Bakhtin 1981, 1986). Vygotsky and Bakhtin should be considered independent inventors of the modern social theory of mind alongside their contemporary, G. H. Mead (1934). Wertsch (1991) stresses that mind is mediated action, and that the resources or devices of mediation are semiotic. Mind, he argues, is socially distributed mediated action.

It is also important to register in these early moments of this effort in theory construction that sociology has something to say about the brain.

Clifford Geertz (1973, 76) has pointed out that the brain is "thoroughly dependent upon cultural resources for its very operation; and those resources are, consequently, not adjuncts to but constituents of, mental activity." Indeed, DeVore (cited in Geertz 1973, 68) has argued that primates literally have "social brains." The evidence for this conjecture in humans has been accumulating in recent years along with a breakdown of the brain/mind/body divisions (e.g., Brothers 1997; Pert 1997).

Our understanding of mentalities has been obstructed by some deeply ingrained assumptions about human beings. One is that affect and cognition are separate and separated phenomena. This division is breaking down (e.g., Zajonc 1980, 1984; Gordon 1985; Damasio 1994; Pert 1997) and will have to be eliminated as part of the process of constructing a sociology of mind. Another assumption is that learning and cognition can be decontextualized. We argue along with other social scientists, by contrast, that learning and cognition are linked to specific settings and contexts: that is, they are indexical. Their long-term efficacies are in fact dependent on contextual recurrence, contextual continuity, and recursive contextualizing. The latter process helps explain the process of generalization without recourse to epistemological mysteries or philosophical conundrums. We live our lives by moving from home or school to home or school, from our home to our neighbor's home, from the schools we attended to the schools our children attend. Contexts repeat, imitate, suggest, overlap, impose and re-impose themselves, shadow and mirror each other, and are linked through simple and complex feedback loops. Indices, like clocks, signs, language, and number, are created to generate shared frames of thought and action. This is the structural and informational basis for the continuities in our sense of self, our memories, and our thoughts. Many of the mysteries of the paranormal and our everyday experiences of déjà vu can be explained by attending to these features of context.

The prevailing theories of theory of mind emphasize development within the individual. From a sociological or anthropological perspective, theory of mind and mind itself are cultural inventions (Astington 1996). Social construction of mind has not been ignored, but it has not been as centrally represented in either mind studies or social robotics. The sociology of mind and thinking has a long and distinguished pedigree, yet it has until recently been virtually invisible in contemporary theories of mind (Valsiner and van der Veer 2000). A renewed interest in mind, brain, consciousness and thinking are one of two quests, the other being the new life evident in searches for God, related to sociology's program for the rejection of transcendence. This is evident in the steady stream of books, articles, lectures, news stories, and television programs crossing today's intellectual landscapes. One of the main features of this literature is that one can see some evidence of a sociological orientation emerging, albeit timidly and fearfully, out of the shadows.

An archaeology or historical study of the ideas of these developments would reveal a "journey to the social" across the entire landscape of intellectual labor. The very fact of the journey to the social reveals the emergence of a new discursive formation, a new episteme. This episteme is new in the sense of a birth or an originating activity, but absolutely new in the scope of its impact. Beginning in the 1840s, the west entered the Age of the Social, an era of worldview changes that will carry well into the twenty-first century and likely beyond before it begins to embody itself in the everyday ecologies and technologies of mind in new global configurations. In this process, what was western and European about the social will get permeated and possibly transformed into a worldview that is less ethnocentric.

On a practical front, think about the difference between thinking about how children learn from a mental process worldview compared to a social process worldview. If you were a teacher or a parent, which worldview would facilitate your ability to encourage, foster, and augment a child's learning? How might a mental process or social process worldview be used to allocate funding for students who are not performing at expected levels? Thus, the theories we adopt help us define, create, and regulate our world. They help us to create the truths that we live by and our notion of how the world should work.

What Can Sociologists Say about Mathematics?

What does it means to say that mathematics is a-social phenomenon, and indeed a-social construction? Consider, for example, the case of non-Euclidean geometry. Euclidean geometry is the geometry you learned in elementary school. It is plane geometry, the geometry of two-dimensional space. This is the geometry of squares, and circles, and triangles. Non-Euclidean geometries (NEGs) are geometries of non-planar surfaces, such as spheres. Euclid (approximately 325–265 BCE) articulated a system of geometry in Alexandria, Egypt. Here's what a distinguished historian of mathematics (Carl Boyer 1968) and an equally distinguished mathematician and historian of mathematics (Dirk Struik 1967) had to say about the emergence of non-Euclidean geometries:

We find a startling case of simultaneity of discovery, for similar notions occurred, during the first third of the nineteenth century, to three men, one German, one Hungarian, and one Russian. (Boyer 1968, 585)

It is remarkable how the new ideas sprang up independently in Gottingen, Budapest, and Kazan, and in the same period after an incubation period of two thousand years. It is also remarkable how they matured partly outside the geographical periphery of the world of mathematical research. (Struik 1967, 167)

There are a number of things about these remarks that are suspect. Words like "startling" and "remarkable" impede rather than facilitate analysis; words like "simultaneity" and "independently" are obstructions to socio-logical insights. A little history goes a long way to eliminating the obscur-ing effects of these words and the descriptions by Boyer and Struik. To begin with, the so-called "incubation period" is really quite active. Non-Euclidean geometries (NEGs) have their origin in problems with Euclid's parallels postulate, noticed by his earliest commentators. A continuous interest in these problems over the next two thousand years is marked by such names as Saccherei (1667–1733), Lambert (1728–1777), Klugel (1739–1812), and Legendre (1752–1833). The first important social fact is that there is a history to, and an historical context for, NEGs.

The German, the Hungarian, and the Russian Boyer refers to were, respectively, Reimann, J. Bolyai, and Lobachevsky. Contrary to the impli-cations in the above quotes, these were not by any means provincial isolates. All had connections to Gauss at the University of Gottingen. This is important because Gauss was writing letters to his friends and col-leagues on NEGs as early as 1799 (to W. Bolyai, followed by letters to Taurinus in 1824, and to Bessel in 1829). Gauss was friends with J. Bolyai's father, and the two had been fellow students at Gottingen. One of their professors was Abraham Kastner (1719–1800), a professor who was an expert on and lectured about the history of the disputes about the paral-lels postulate.

The parallels postulate was also being studied at the university in Marbourg. Marbourg professors such as F. K. Schweikart (1780–1859) and A. L. Gerling (1788–1864) were more or less directly connected to Gottingen and Gauss. Gerling had studied with Gauss, and Schweikart's nephew Taurinus (1794–1874) had studied at Gottingen.

The supposed provincialism or regional isolation of the creators of NEGs does not stand up to the most cursory scrutiny. The Bolyais lived in an outlying Hungarian town, but both father and son devoted a great deal of time to the parallels postulate. J. Bolyai had developed his ideas on NEGs as early as 1823, and while his father was dismayed by his son's fascination with these intractable problems, Johann's "The Science of Absolute Space" appeared in a book Wolfgang published in 1832/33.

At first glance, Lobachevsky seems to qualify as a provincial isolate. However, the university he studied at in Kazan was staffed by distinguished German professors. His teacher J. M. Bartels (1769–1836) had studied with Kastner at Gottingen, and was one of Gauss's teachers. And Reimann (1826–1866) studied at Gottingen as Gauss's student.

There are two points to make about this history. First, there is clearly a social network at the center of the development of the NEGs. The lesson for sociologists of knowledge is, given isolates, or cases attributed to genius, look for a network. Second, the myth that mathematical systems like NEGs are pure creations of the human mind with no connection to real world problems is readily dismissed. It's true that Lobachevsky, for example, referred to his geometry at first as "imaginary." He believed that geometrical ideas had to be verified like other physical laws, but in the beginning he did not have a clear physical perspective on NEGs. NEGs were in fact created to solve problems that had arisen in mathematics and in studies of physical phenomena.

The case of NEGs illustrates the fallacies embedded in traditional ways of telling stories (histories) of mathematics, stories that reinforce assumptions about Platonic realms and objects. This is just one case, but it is paradigmatic. In the words of Simon Stevin (c. 1548–1620), the late Renaissance scholar, "Nothing is the mystery it appears to be" (this was the motto he fixed to his coat of arms). It illustrates how "facts" reflect cultural values. The stories told about supposedly isolated mathematicians reinforce both the idea of "discovery" of the *truth* as well the myth of the lone genius. Social stories, ones that focus on networks and connections, provide far better models that actually explain both the production of "new" ideas and the continuity of "old" ideas, and for tracking change, conflict, and ideas about proof, reality, and knowledge present in various cultural formations.

Conclusion

What counts as proof, as evidence, as real, varies by culture and even within cultures as people are tied to networks and worldviews and differences in social position, gender, race, and class. What counts as science, as reliable and useful knowledge, and the validity that science might have in relation to other claims of knowledge is similarly varied, and subject to dispute. What counts as the "best" technology can never be a straightforward matter of technical criteria, because social processes are necessary to decide which features of a technology are most important in the inevitable selections among trade-offs in design, manufacture, and use. These propositions seem disarmingly simple, but their complexity and power to engender insights and their implications are rich resources for research and critical thinking. The idea that science and technology, logic and objectivity, and the very notion of truth are socially manufactured is an admittedly difficult idea, but only because the sociological principles it rests on are not widely taught. It is an idea too that is a tough challenge to existing systems of authority based on unquestioning acceptance of expertise. As we approach the end of this chapter, we find ourselves reiterating the central tenet of science and technology studies; science, technology, logic, and mathematics are socially constructed. This is often misinterpreted as "relativism" or "anti-realism." In fact, social construction is the only way we have of manufacturing our cultures, our truths, our falsehoods. It is not social construction that realists have to fear, but rather absolutism, universalism, imperialism, and colonialism and other "isms" representing barriers to inquiry.

Notes

1 This debate has recently been reopened with the identification of Sedna, a rocky, icy, planetary body somewhat smaller than Pluto in the outer reaches of the solar system. Rethinking the classification of Pluto is again on the agenda at the IAU (Briggs 2004).

2 See for example, the recent anthologies by Lederman et al. (2001), and Wyer et al. (2001).

3 The rationalists are the intellectual descendants of Descartes and Leibniz; the empiricists follow in the footsteps of Locke, Berkeley, and Hume; Watson is the source of the behaviorist challenge to the radical empiricists; the ethologists Lorenz, Tinbergen, and von Frisch offered a different kind of challenge to the behaviorists. The Kantian counterpoint to empiricism shows up in Fodor's conception of the mind as an entity possessing organizing capacities and an innate "language of thought."

Further Reading

Cole, S. (2001). *Suspect Identities: A History of Fingerprinting and Criminal Identification*. Cambridge, Harvard University Press.

Fleck, L. (1979/1935). *Genesis and Development of a Scientific Fact*, tr. F. Bradley and T. J. Trenn. Chicago, University of Chicago Press.

Kay, L. E. (2000). *Who Wrote the Book of Life? A History of the Genetic Code*. Stanford, Stanford University Press.

Oudshoorn, N. (2003). *The Male Pill: A Biography of a Technology in the Making*. Durham and London, Duke University Press.

3

The Dance of Truth

Our goal in this chapter is to develop an appreciation of technoscience as a social institution and to comprehend more about it by looking at its similarities and differences compared to magic and religion and the law as institutions. This will aid in understanding the idea that institutions are powerful, productive, and also limiting and dangerous. We will be introducing a set of ideas about truth, knowledge, and objectivity that are hardly new, but nonetheless still under-appreciated.

Before we do this, however, we want to return to an earlier concept. In chapters 1 and 2 we discussed culture as norms, practices, language, material culture, and symbols. So could we call science a culture? To identify a system as a culture we would need to show that it has these features. There are norms and expectations about behavior in science. Scientists are expected to be truthful and honest in their work and to report it in the same manner. When scientists transgress this norm, they risk losing their reputation, funding, and position. Within science, dishonesty may be the gravest of sins while truthfulness is the most esteemed virtue.

Technoscience reproduces and maintains itself by training and educating scientists. If individuals want to become scientists, they need to attend college for four years, do a post-graduate degree or two, and then work as a post-doctorate fellow/junior researcher until eventually they publish enough to establish themselves as senior scientists. In some cases, collaboration and funding are key factors in moving up the ranks.

Science has a material culture that is typically composed of a laboratory with special equipment, measuring devices, and recording devices. Others, observational sciences, such as geology, ecology, or astronomy, have a different configuration because their sites are not constrained by four walls. For each field of science the equipment may vary, but the intended purpose of the equipment is to measure, record, and describe. Also recall the stereotypical image of a scientist and the way the white lab coat and glasses symbolize "science."

Technoscience has symbols (the white lab coat is one) and a unique language that unites scientists into a community and places them within a-social system. These symbols also carry ideas that are shared by a particular culture. For the society at large, a white lab coat symbolizes the purity of science, the intelligence and devotion of the scientist to truth, and the difference between scientists and tinkerers in garages. For the science community the white coat symbolizes technicians working in a very specialized lab with a particular focus. It also reflects ideas about keeping clothes clean while in the laboratory, creates a boundary between the laboratory and everyday life, and serves as a marker of status, position, and activity.

Mathematics is a unique language of symbols used by science to unite its practitioners, to exclude the uninitiated, and carry information. When children are identified as understanding the mystery of numbers, they are encouraged to study more mathematics and science. A child with a weaker initial grasp of numbers and their meaning is often channeled in directions away from disciplines that depend on mathematics. Because fewer are initiated into the mysteries of mathematics, it carries more power and prestige that translates into physics being the "hardest" science of all. Biology is not quite so "hard" because it is thought to be less mathematical. Stratification of individuals and disciplines is played out with and through the symbols named mathematics.

We are beginning to see how science can be thought of, analyzed, and understood as a culture. We can now begin to answer some of the big and small questions of science studies. For example, why are there more Ph.D.'s awarded to women in biology compared to physics? Why is it that Asians have entered science and engineering at a greater rate than Africans or Mexicans? Do scientists cheat? How is an idea or a fact born, nurtured, and established? In addition to thinking and analyzing technoscience as a culture, we can also analyze and describe it as a social institution.

Science and Technology as Social Institutions

All societies have social institutions that organize people and ideas around certain goals that the society values. Typical social institutions are the family for reproducing and maintaining membership in the society; education for preparing the young to participate in the society; economic systems for organizing the activities of production and consumption; politics for organizing power and providing mechanisms to establish and sustain stability and order; and religion for providing a collective way of knowing about the certainties and uncertainties of life, death, and existence, and for institutionalizing a moral order. This picture of institutional life emphasizes the basic requirements for the efficient functioning of a society. A more complex analysis would demonstrate the slippages and anti- or alternative institutions that are the seedbeds of conflict and social change.

Historically, science and technology have not always been included in social institutional analysis. However, this is changing as contemporary society and science and technology or *technoscience* become more intertwined and not simply a piece of a simple institutional order. The values and ideals of technoscience are becoming more and more important for defining the core aspects of life and who we are.

Social institutions are sets of durable social relations. All institutions are organized around social roles, rules of conduct, forms of organization, social and material practices, and often specific languages or discourses. What, then, does it mean to say that science is a social institution? First, we have to distinguish science as a generalized activity found in all human societies from modern science. Second, we have to understand modern science as a social institution embedded in a modern industrial-technological society. Modern science is a social institution with its own set of social roles, including technicians, researchers, teachers, and students. It also has rules, some of which are explicit (don't plagiarize) and some more or less implicit (e.g., publish frequently). Scientific values, such as objectivity or neutrality, or prestige, are part of the normative or rule structure of science. Robert Merton (1973) articulated four basic norms of science: disinterestedness, universalism, communalism, and organized skepticism. Many scientists believe this is how science does or should work. Science includes experiments or field studies, observations, and communication practices (including conferences, publications, and even electronic listserves). Science can appear in different organizations, such as universities, research institutes, and other communities of different sizes. And of course science has its own specialized languages, reflecting specific rules and values as well as specific disciplinary knowledges. Sometimes this specialized language seems to be jargon unintelligible to outsiders and the lay public, but it enables specialists to communicate precisely and efficiently. Discourses and practices taken together are the bases of subcultural formations that shape the assumptions about what is real, what is possible, and what is desirable for a social group.

When we study technoscience coupled with studies of organizational cultures and complex organizations in an institutional framework, we can begin to see the influence that routine, myth, and ritual have as necessary parts of producing science and technology. One common myth of technology is that it makes life easier. It can certainly alter the amount of energy expended directly by a person to do a certain task. For example, simply touching a button to play music is easier than searching through a pile of records, placing a record on the record player, and carefully aligning the needle to hear music. Using a record player is easier than finding and shaping the materials needed to make a drum, flute or guitar and then learning to play it to produce music. Touching the play button on the computer is an easier way to create music than was possible one hundred years ago. However, what about the hours, sometimes weeks of frustration that might go into setting up the computer? Do we include the time and energy and human labor in general that go into designing the software and hardware when we make comparative judgments about "ease" and "progress?" The women who work in maquilas or other global sweatshops do not find their work easy as they assemble computer components for very low wages. Where do we stop and start the analysis for determining whether technology makes life easier or not? And how much of our use, design and creation of technology depends upon this core belief that technology makes life easier?

Ritual is not often associated with science but with religion. However, when we observe technscience in practice, we find ritual and routine infuses practice and discourse. One reason for this is the scientific method and the repeatability values that are embedded in it. Technique becomes ritual as it is repeated. Examples of common rituals in science include what the surgeon does before going into surgery, what the technicians do to get equipment to run properly or what academic scientists do in the ritual performances of reading papers at conferences. What appear at first to be simply habitual or regularized patterns of behavior turn out on closer examples to be rituals that sustain social solidarity.

Many look at the social and cultural components we are drawing attention to as detracting from science and technology and argue that they have no apparent function except the comfort of routine. However, they are no more than manifestations of a complex activity, and all forms of organization have their everyday working systems and pathologies, functions and dysfunctions. The dysfunctions are not "side effects" or simply a reflection of bad behavior by individuals, but the hidden side of complex organizations and practices and are produced by the same mechanisms that produce the obvious results and the "good things" in organizations (see Vaughan 1999). For example, in the 2002 stock market ethics scandals, the mechanism of stock options, which provides incentives for upper management to increase their productivity, also provides incentives for overstating the value of a firm and promotes short-term thinking about the price of stocks rather than the long-term productivity, profits, and dividends of the company itself. The idea that we want to get across is that all social institutions are productive, yet fallible. Science, as a social institution, has developed mechanisms, such as peer review, which help to increase the reliability of knowledge claims. However, peer review can also lead to difficulties for newcomers or outsiders who want to contribute to science and are prevented by the very process designed to verify science. It can be used unfairly or even unethically when people review materials without rigor because of bias (pro or con) toward an author or idea. It is even possible that a reviewer might steal ideas from pre-published materials.

So ritual and routine are part of technoscience, and subject to the same possibilities of flaws at the same time that the routines produce reliable results. Clearly, science as a generalized activity has durable social relations with sets of social roles, rules, norms and practices. What happens if we substitute technology for science into the previous phrase? Is it still a viable phrase that can teach us about technology and society?

Technologies are material products of human activity, including tools, toys, artifacts and artworks, "low" and "high" technology, as well as knowledge about how these products work and especially how to make and use them. A good example is a fire bucket used to carry water to put out a fire. It is a tool. When it is put in a museum it is an artifact. In both cases it is a technology. What is subtler is to consider the fire brigade as a technology: knowing how to organize people into lines to move a fire bucket from its source at the village well to the scene of the fire is also a technology. In that way, an organizational strategy can serve as a technology along with typical items like computers. There is a tendency in the United States and other nations with a lot of "advanced" technologies to overlook "low-tech" solutions. People in such nations tend to overlook the importance of infrastructure (e.g., roads, ditches, bridges, and sewers)

– at least until these systems fail. They often overlook items in domestic life, such as dishes or baby bottles. These items are parts of a shared material culture and important politically and economically, and provide the material bases of important institutions themselves. Emphasis on "high-tech" systems reinforces the idea of progressive technological change.

Wiebe Bijker (1995, 2001) has written extensively on the social construction of technological systems in ways compatible with our thinking about technoscience in institutional terms. He discusses both the material durability of technologies as well as their semiotic power, the accumulated meaning that can both make technologies seem inflexible or provide opportunities for change. In Western societies, people tend to see science and technology as organized around common physical principles and shared goals of control. Sociology uses the same theoretical toolkit to understand and explain science and technology. These are among the reasons for using the idea of *technoscience* to describe knowledge systems, thus emphasizing our understanding that it is impossible to separate knowledge from its applications. We do recognize that engineering and technology are not the same as science, but this is because the two fields are in different communities of practice and different institutional locations. And, to anticipate a later discussion, we should see much contemporary technoscientific work as part of a larger system of "planned innovation." This may seem like a contradiction in terms because many imagine the invention process as spontaneous and uncontrollable. But it underscores that contemporary research is conducted on basic physical systems, such as semiconductor quantum effects, but that it is directed toward the social goals of various actors, such as improving hardware efficiency for computers.

The Dance of Magic, Science, and Religion

Karl Marx (1964, 53) wrote that the criticism of religion is the beginning of all criticism. If we do not have a critical perspective on religion, if we cannot theorize about religion, if we cannot explain religion as a social construction, we will be hampered in our efforts to understand the social nature of science. The reason for this is that the barriers to inquiry that have traditionally surrounded religion are barriers to inquiry in general. They are barriers to seeing the social in inquiry, barriers that sustain the idea that there are transcendental and supernatural worlds. The idea of the supernatural has been worn down by the successes of the natural sciences. It will take comparable successes in the social sciences to finally undo the seductions of the transcendental. As we undertake this journey to eliminate the transcendental, we will find ourselves contending with magicians and magic or being thought of as magicians.

Like Marx, many of the nineteenth-century founders of social theory were concerned with understanding and explaining religious beliefs and institutions. This was an age of doubters, debates over the religious implications of new discoveries and theories in geology and biology, and the crystallization of thousands of years of speculation into a systematic social theory of religion. You may find this description surprising because you believe that there is no way to prove or disprove religion or God. Or perhaps you believe that because religion is a matter of faith – and individual faith at that – God and religion are outside the bounds of scientific explanation. This and similar ideas are so powerful that even critics and theorists of religion as a-social phenomenon are still hedging their bets. While undermining all possibility of reasonable grounds for believing in God, many writers caution that no matter the strength of their criticisms or arguments, believers really have no reason to alter their beliefs or practices. We are not adopting that line here. It is our contention that in our intellectual community (thought collective, objectivity community) we know there is no God with the same tentativeness and corrigibility that accompanies our critical certainty that the earth is not flat.

One of the most important tools we have for understanding and explaining our ways of living and thinking is the comparative method. Using the comparative method demonstrates that different types of societies generate different types of gods and religions. The nature and extent of the division of labor, the degree of social differentiation, the type of stratification system, and social changes (within and across societies) are some of the factors that determine the moral order of a society. Religion is in fact just one way of systematizing, organizing, rationalizing, and institutionalizing a moral order. All societies, all individuals, must have norms, values, and beliefs about right and wrong behavior, good and bad behavior. Most, if not all, societies have traditionally rationalized their moral order through some form of religion. As we explore the social nature of religion, it will be useful to keep in mind the proposition that moral order is a primary component of religion.

It seems so natural to think of magic, science, and religion as somehow tied together. In our everyday lives, science can sometimes seem to be magical. There used to be a physics lecture at one of our universities that was called "The Magic Show." Magic as we know it through the performances of contemporary hi-tech magicians, such as David Copperfield, can seem to be as much about the art of illusion as it is about the science and technology of illusion. Most of us probably have some vague ideas about a connection between magic and religion, but we are not likely to have thought through and articulated the nature of that connection. Our everyday intuitions about the relationships between magic, science, and religion may reflect a long-standing assumption among sociologists, anthropologists, and other scholars that the three form "a three cornered constellation" (Malinowski 1954/1948).

The study of magic, science, and religion gets under way in a systematic way in the nineteenth century with the crystallization and development of the social sciences and social theory. By 1925, Bronislaw Malinowski, one of the founders of modern anthropology, was able to look back on nineteenth-century students of society and culture and find a variety of perspectives on magic, science, and religion. Those earlier thinkers had focused on explanation, projection, and ritual in trying to understand the three-cornered constellation. Malinowski (1954/1948) came to see that magic, science, and religion are linked because they are rooted in and help to define the two domains of the sacred and the profane found in all human societies. Magic and religion, on this view, are ingredients of the sacred domain; science occupies the profane domain.

From a variety of perspectives, some individuals see a conflict within the three-cornered constellation (Nader 1996). Magic, they claim, is related to science and religion in the same way. Magic is perceived as disreputable and unacceptable in contrast to the socially acceptable activities of science and religion. If the three-cornered constellation becomes science, technology, and society, it takes on a new significance. This new constellation is much more complicated than earlier thinkers imagined. If the constellation or triadic relationship is indeed a cultural universal, then we should find it operating in modern science and technology as well as in modern religion. It should be operating in magic too, both the everyday magic of the illusionists and in the institutional magic that is linked with science and religion.

In the ancient Palestinian period out of which the Jesus stories of the New Testament emerged, Jesus was viewed as a "magician." The term "magician" is understood by students of that period to mean, among other things, "miracle worker." The term, however, covered a range of social roles from street preachers to the teachers of emperors (Smith 1978, rpt. 1987). Within this range would be found men who were associated with the reigning religious traditions, the magi. There were traditions that vilified the magi, including Jesus, and others that revered them. There were even magicians among the gods, including Circe and Isis.

By the late twentieth century, it seemed clear that science and religion were more similar than dissimilar. Our focus here is on an institutionalized complex of overlapping practices and discourses. Magic has not received the attention of late that it did in earlier times, especially in terms of science studies. There seems to be a renewed interest in magic; probably because we are getting comfortable with the fact that science and religion are at least in part about desire and emotions. Magic, science, and religion are our institutional names for the ways in which our society orders desire, reason, and humility or faith (Nader 1996). In a more general sense, the three-cornered constellation is the locus of our controversies over relativistic versus realistic perspectives on the world.

How do contemporary scholars view magic? Anthropologists tend to define it as a practice or a formula designed to produce a desired outcome. The desired outcome might be personal gain, warding off harm, or harming one's enemy. It soon becomes apparent when we try to get to a definition that it is not always easy to distinguish magic and religion, magic and prayer, or magic and medicine (O'Keefe 1982).

Religions and gods come in many varieties. It is this variety revealed by comparative analysis that can first provoke doubt among believers. This can lead to considering the possibility that religion is social, and not transcendental or supernatural. The forms that religions and gods take vary with variations in the division of labor, social differentiation and stratification, and political economy. Locality and culture give form to the gods. We find gods of war in warring societies, fertility gods dominating in agricultural societies, and gods or goddesses predominant in societies that are respectively male dominated or female dominated (or else egalitarian). Monotheism is associated with societies that have at least three levels of sovereignty (e.g., clan, city, empire). Class-dominated societies produce polytheism, the belief in multiple gods. Ancestor worship is associated with extended family structures, and reincarnation with intense face-to-face village communities. The conception of a Supreme Creator interested in human morality is rare among hunter-gatherers, horticulturalists, and fishing societies. The idea is, on the other hand, common in agrarian and herding societies. Such correlations show up more or less clearly in historical perspective. But religious institutions, belief systems, and gods can mix in very intricate but still comprehensible ways in more complex societies.

Sociologists and anthropologists can identify religious practices and institutions by looking for a sacred and profane division of the world. In the profane world, life is practical and matter of fact. In the sacred realm, behavior is strictly controlled, serious, respectful, and organized with great care. Rituals and worship support social solidarity and aid in preserving traditions. This idea, which we can express in the formula, God equals society, or the community, was first discussed in sociological detail by Emile Durkheim.

Durkheim's study of aborigine totemism in Australia led him to conclude that the sacred objects people worship are surrogates for their communities. When we worship God, in other words, we are worshipping society. Thus, the referents for religious and spiritual beliefs are to be found in our earthly lives, our collective actions, and our social interactions. Transcendental and supernatural referents are illusions.

Religious experience derives from ritual behaviors in the earliest human societies. These periodic collective activities gave rise to the sort of excitement we still experience today when we enter a concert hall or get caught up in crowd behavior. What we feel and how intensely we feel depends on whether we are in a group listening to chamber music or at a rock and roll concert, as well as what our musical preferences are. We can also experience a related kind of excitement when we get into the rhythm of a group of people at a party or at an intimate dinner. The *group* was not a sufficiently demarcated thing or object in the worlds of our earliest ancestors. Individuals, as fully self-conscious egos, were also not clearly separated from the collective. Over many centuries and even millennia, objects were selected as symbols of the feelings or emotions generated during solidarity gatherings. Raw group activities slowly evolved into rituals and then into rites oriented around such objects. Collections of rites, myths, and beliefs clustered around sacred objects developed into cults, interrelated and rationalized cults became religions. This process, schematized here, unfolded over tens of thousands of years.

Individual emotional experiences during rituals are in fact social phenomena and cannot by themselves explain the origins of religion. A combination of limited experience, the nature of language and grammar and their volatile potential for reification through naming, the need to classify and categorize (the boundary imperative), and the capacity for generalization and abstraction all helped to generate mistaken beliefs about the referents for the affective consequences of ritual. Rituals and the boundary imperative are the roots of ideas about supernatural and transcendental realms.

Consider an example of how the processes we have been discussing operate in the Hindu tradition. Fire is considered a god, or the body of a real god, in the ancient Hindu Vedas. This idea evolves over time into the concept of an eternal god of all fires. Eventually, this concept is stabilized by a cult of this god of fire. In general, the emergence of cult specialists signals the coming of a religion. Organizational activities that systematize ideas about the gods are reflected in generalization, abstraction, symbolization, and reification.

In earlier societies, people are not as alienated (distanced) from the basic rites that generate gods and religions. In such societies, some people might be aware at some level that the gods are created, sustained, nourished, and rejuvenated by the rites, that people literally manufacture the gods. At our stage of cultural development, the concepts and emotions necessary for consolidating that sort of awareness are not available to most of us. It is mostly the political, religious and intellectual classes who now have access to these resources, and more because of their historical consciousness than because of any direct experiential access. The message will go out from these classes, especially from the intellectual classes, that we humans ourselves manufacture the gods and religion. This message will not reach most people; indeed, it will not even appear in their schoolbooks. It is no longer transparent that religion is related to solidarity activities. Modern humans are deeply alienated from this relationship, and will resist explanatory efforts by philosophers, sociologists, and other intellectuals.

People will be most acutely aware of their role in constructing gods as symbols in the earliest moments of the emergence of a new society or political constitution. In these moments, they can see themselves designing social orders, creating gods to symbolize and celebrate their new moral order. In the wake of this celebratory period, religions become moral prescriptions for current and future generations. The moral order is rationalized in religious organizations and beliefs on behalf of the ruling powers. As alienation sets in, people forget that they made the gods.

Moral orders systematize ideas about right and wrong, and good and bad. They are intrinsically social and regulative. Moral orders seem to be and are experienced as if they are outside forces. But they are group forces, not transcendental ones. We are social animals, and perhaps the most social of the animals. Being human means desiring to belong to the group. This desire can vary, notably in complex societies where competing and conflicting social forces can cause some of us to be resistant to belonging and to become alienated from group behavior. This can affect our degree of social integration and lead to lives in the margins of society. Some level of marginality seems to be required for creative, critical living. Carried to extremes, it can also be a cause of deviance, mental illness, and sociopathic behaviors.

Moral behavior is a given for well-integrated members of society. Punishment, banishment, imprisonment, and death are some of the ways societies try to control morality. In the earliest human societies, morality is built up from the rules of behavior necessary for survival. As societies become more complex and divisions of labor and systems of stratification emerge, moral order increasingly reflects the interests and values of the ruling powers and contributes to sustaining prevailing modes of power and domination. Multiple group memberships, group conflicts, and patterns of joining and leaving groups are some of the sources of moral conflict in the individual person. Religion systematized the moral order in terms of the rules of right and wrong appropriate to people's social class or station in life.

In societies with less complexity than the agrarian level, political and religious authorities are not institutionally separated. That separation, in the form of new institutional linkages, begins with the emergence of agrarian, commercial, and cosmopolitan societies. If the idea of an alliance between politics and religion sounds strange to you, consider the roots of religious marriage ceremonies. Where does the priest's, minister's, or rabbi's power to marry come from? The marriage license is a clue. State power stands beside you at the altar or before the justice of the peace. We need just a few more analytical steps to uncover religion's function in legitimating political power and supporting a moral order that reflects the state's interests. The historian of the rise and fall of Rome, Edward Gibbon (1776–1788, rpt. 1993), pointed out that the various religions that flourished in that era were considered equally true by the masses, equally false by the philosophers, and equally useful by the magistrates.

Organized, institutionally distinct religion developed in step with military and political institutions. They kept enough autonomy to support the distinction between "this worldly" and "other worldly" realms. New pathways to salvation arose as the world religions of Judaism, Christianity, Islam, Confucianism, and Buddhism emerged and crystallized in the cosmopolitan centers of the ancient world. These were to different degrees and in different ways transcendental religions. Health, wealth, and military success were the signs of spiritual well-being in traditional societies. Spiritual transgressions and the like were the causes of bad luck.

The after-life became a solution to the trials and tribulations of this life in Christianity and Islam. In the ethical religions (e.g., Confucianism), the stress was on "right behavior" in the everyday world and not on salvation in a future real (as opposed to metaphorical) heaven.

The legends about gods, messiahs, wonderworkers, magicians, and heroes come out of the same basic molds. The lives of Apollonius, Jesus, Simon Magus, Gregory the Wonder Worker, and from more modern and recent times Nostradamus, Rasputin, Aleister Crowley, and Eduardo the Peruvian healer are variations on some common themes. Their stories (often reformulated over time and after their deaths to comply with the stock legendary features of the magus' or heroes' – including mythic heroes' – lives) have some or all of the following features: the divine origin and miraculous birth, the annunciation and nativity portents, the menace to the future magus during infancy, the initiation, the trial of spiritual strength (temptation resisted after a long solitary fast), miracles, the sacrificial feast, trial and death (by crucifixion, for example), the disappearance of the body, and descent into hell, resurrection, and ascension. The magus themes are also found in tales about outstanding ancient figures. It is not too far from legend and myth to the idealistic portrayals of scientists and philosophers. Plato is a good example. The famous philosopher was transformed into a divine figure by his nephew Seusippus during his eulogy for his dead uncle. Seusippus claimed that Plato was the son of the god Apollo's alliance with Plato's mother. This may have been only a "manner of speaking," but as a metaphor illustrating the origin's of Plato's talents, it nonetheless fed into beliefs about personages such as Jesus, who is still believed by millions of people around the world to be the son of God. The relationship between the hagiographies of scientists from Newton to Einstein, and from Archimedes to Galileo and their lived lives are filled with stock images of the "pure" scientist, which continue to reflect messianic and heroic narratives.

It was already clear to some thinkers and philosophers in the ancient world that human beings create religion and the gods. This idea crystallized into one of the great discoveries in social science in the course of the nineteenth century. The evidence for the discovery of God had been accumulating for centuries. On the basis of proof by conciliation of inductions and ensemble of probabilities, we can now dismiss the claim that it is impossible to *know* whether there is or isn't a God. We are now in the realm of a good sociology of knowledge problem. The "we can never know" claim is at worst an effort to prevent this discovery from reaching "innocent" and "vulnerable" minds. It is also based on ignorance of the scientific grounds of social theory, or an unwillingness to acknowledge even the possibility of such grounds.

The idea fashionable in intellectual circles that only the physical and natural sciences can lead us to knowledge of God is curious in two ways. In the first place, this erases the realities and potentials of the social sciences as explanatory and discovery sciences. In the second place, while some of us might be able to mount an argument against the existence of

God based on evidence from the physical and natural sciences, it is interesting that in the public arena physical and natural scientists are most visible as God "provers."

Once social science as science is admitted into the arena of inquiry on gods and religion, order is rather quickly achieved without resorting to transcendental and supernatural notions, and without retreating to metaphysical agnosticisms. In recent decades, efforts to systematize and formalize the evidence on the social bases of gods and religions have been undertaken. It has been easier in recent years to find very good college textbooks on the sociology and anthropology of religion. It is still fairly common, however, to leave the question of whether there is or isn't a God open to individual interpretation even while page after page the author destroys the foundations of belief. The reasons for this are fairly obvious.

Marx, who is so well known for his comment on religion as the opiate of the masses, was also astute enough and realistic enough to point out that religion is not simply invented by scoundrel priests and rulers. Religion is a search for relief from human suffering and the quest for comfort and security in an alien universe. One of the features of so-called democratic societies is a tolerance for religious differences. But there are invidious ways in which such tolerance is used to oppose inquiry. We certainly don't believe we have the right to impose our views on others or that our authority automatically overrides the authority of alternative viewpoints. But the ethos of our science, of our theory, requires that we adopt, and present, and defend perspectives consistent with the achievements of our fields of inquiry. Marx did not look forward to an atheist or agnostic society, a society that needs to deny God or leave the question of God's existence open, but rather to a society so transformed that the question of God would not exist. Such a society, like all societies before it, will of necessity have to provide the functional equivalent of the gods and religion, as we have known them, that is, a moral order. What, then, is the nature of the truths we hold as social theorists of religion, science, or of society in general?

What Is Truth?

If you are familiar with the New Testament, the title of this section may remind you of the following dialogue between Jesus and Pontius Pilate (John 18: 37–39):

"You are a king, then!" said Pilate.

Jesus answered, "You are right in saying I am a king. In fact, for this reason I was born, and for this I came into the world, to testify to the truth. Everyone on the side of truth listens to me."

"What is truth?" Pilate asked. With this he went out again to the Jews and said, "I find no basis for a charge against him. But it is your custom for me to release to you one prisoner at the time of the Passover. Do you want me to release 'the king of the Jews'?"

Now Pilate might have meant, "What difference does truth make, what does it matter?" or he might have had in mind something about the difficulty of finding out the truth. The truth quest is an ancient sport, and ancient thinkers did not miss the problematic nature of the object of the quest. By the late nineteenth century, philosophers were getting quite sophisticated about the problem of the truth quest. No one offered more profound criticisms of and insights into this quest than Friedrich Nietzsche. It is not our intention to draw you into the subtleties of these criticisms and insights. What we want to do is draw your attention to the fact that telling the truth and searching for the truth are very tricky and very risky enterprises.

Science and truth speak in the interests of and are spoken by the powerful. The Enlightenment held that "the truth will set you free," that knowledge is power. Modern thinkers recognize that power is knowledge.

The development of postmodernism has made the truth business even trickier and more risky than ever. This is as good a time as any to make clear that science studies researchers believe in reality and assume it is possible to tell the truth. The trick is that our understanding of reality and truth are not what they were a century ago, or five centuries ago, or two thousand years ago. Postmodernism has changed all that. It has led some very thoughtful people down dead-end relativistic paths, put some on one road or another of spiritual or mystical enlightenment, and led some of us to construct new ways of telling the truth. This "new" way of telling the truth is mixed up with the beginnings of the social sciences in the nineteenth century.

One of the truths that we can still tell today in a way that Marx and others would recognize is that there is no God. Another truth is that religion is about moral order and social solidarity, and not about a real heaven or a real hell. Telling the truth about religion is the starting point for telling the truth in general. Religion and God have been protected from our kind of inquiry for so long that it was necessary now to give it more attention than usual in a book about science and technology studies.

Postmodernism has not only taught us to be, at the very least, cautious about "telling the truth," it has also in a way made it possible to tell the truth more realistically. From the beginnings of the sociology of knowledge in the 1930s to the emergence of science studies in the late 1960s, the specter of relativism has haunted those who have sought an objective path to truth. Postmodernism has enhanced the power of that specter. History has many more or less transparent and painful lessons to teach us about searching for and hanging onto truths for dear life. So what are we to do, those of us who refuse to let relativisms and nihilisms (refusal to believe in anything) corrupt our inquiries?

Nietzsche (1968/1888, 50), who carried the burden of nihilism into his inquiries, thought of truth as one of those big words (like *peace* and *justice*) that have "value only in a fight, on flags, *not* as realities but as *showy words* for something quite different (indeed, opposite!)." What is truth, then? "Inertia; that hypothesis which gives rise to commitment; smallest expenditure of spiritual force, etc." Just as there are many kinds of eyes, there are many kinds of "truths." Therefore, "there is no truth" (Nietzsche 1968/1888, 291). Consider now that Nietzsche also said that the more eyes we have the more objective we are! What's going on here?

The problem Nietzsche is addressing in his interrogation of truth is two-fold. Clearly we know from our collective historical experience that it is easy for truths to become Truths. The capital "T" represents the transformation of truth into a God-like thing. When we commit ourselves to "Truths," we inevitably run up against the fact that we live in a dynamic world. Over-committing to Truths will sooner or later get us into trouble by impeding our ability to keep track of changes, notice new things, adapt to changes in related truths. The world has shown itself to be more complex than we can grasp at any given time. We need to remain flexible as inquirers, and Truths, like Gods make us stiff and block inquiry. The other reason we need to be wary of Truths is expressed in the following conversation between the worldly and evil Vorbis and the innocent novice Brutha in Terry Pratchett's novel, *Small Gods* (1992, 273–274):

"We spoke once" [Vorbis says], "did we not, of the nature of reality?"
"Yes."
"And about how often what is perceived is not that which is fundamentally true?"
"Yes."
Another pause . . .

"I am sure you have confused memories of our wanderings in the wilderness."

"No."

"It is only to be expected. The sun, the thirst, the hunger . . ."

"No, lord. My memory does not confuse readily."

"Oh, yes, I recall."

"So do I, lord."

Vorbis turned his head slightly, looking sidelong at Brutha as if he was trying to hide behind his own face.

"In the desert, the Great God Om spoke to me."

"Yes, lord. He did. Every day."

"You have a mighty if simple faith, Brutha. When it comes to people, I am a great judge."

"Yes, lord. Lord?"

"Yes, my Brutha?"

"Nhumrod said you led me through the desert, lord."

"Remember what I said about fundamental truth, Brutha? Of course you do. There was a physical desert, indeed, but also a desert of the soul. My God led me, and I led you."

"Ah. Yes, I see."

Vorbis's power is that everything he does and says is right and true with a circular or tautological definition. We can call this the Big Brother problem. Truth can and does get tied up with issues of power. Now in the end, the important point for us is that all of the subtleties and complexities of the philosophy of truth that Nietzsche struggles with do not stop him from inquiring into the "real." Now you will rightly say that surely there is a philosophy of the real that parallels the philosophy of truth; and indeed, that parallels the philosophy of every word or concept that might demonstrate our interest in how the world really works. So the problem here might be that we have gotten ourselves muddled in philosophy. Suppose we leave Nietzsche, rightly unintimidated by any of this philosophy, to pursue his inquiries and his thinking, and turn to a more sociological perspective on truth.

Ludwik Fleck, a physician–scholar writing in the 1930s, and heavily influenced by some of the classical social theorists, developed an STS approach to scientific facts. He anticipated the founders of STS and often with greater sociological acumen. Let's begin straight off with Fleck's (1979/1935, 100) definition of truth as a "stylized solution." Truth, he wrote, is a particular historical event constrained by the stylized ways of thinking peculiar to a given network of thinkers. It is neither relative nor subjective. What this boils down to, then, is that a given fact can never be true for one person and false for another:

> If A and B belong to the same thought collective, the thought will be either true or false for both. But if they belong to different thought collectives, it will just not be the same thought. It must either be unclear to, or be understood differently by, one of them.

The most important lesson to be learned from becoming sensitized to the problems that get in the way of telling the truth is that inquiry and thinking are not interfered with. After we have been led through the dark and gloomy labyrinth of truth as ideology by Nietzsche and others, and enlightened about the historicity of truth by Fleck and his followers, we still want to know what to do about lies and anti-factual claims. We still want to know how to tell the truth. We want to be able to know what is really going on when someone tells us a child is six years old and someone else tells us the child is ten. Assuming there are records and parents available, we should be able to sort things out. And in fact the child cannot be both six and ten at the same time. The earth cannot be both flat and an oblate spheroid wobbling in precession – at least not for us. Now none of this sorting out has to be simple or straightforward for it to be possible.

The way out of all the philosophically constructed barriers to telling the truth is to drop the "individuated subject" assumption. If knowledge and truth telling are functions of the individuated subject assumed in post-Cartesian philosophy, all the classical and postmodern arguments against being able to tell the truth will stand. We have seen all along in this book, however, that the idea of an isolated subject, brain, or mind is a failure of the sociological imagination. We are now prepared to learn how to tell the truth after postmodernism and in the shadow of Nietzsche. We take our lead here from sociologist Dorothy Smith (1999, 127):

> Knowledge, and hence the possibility of telling the truth and of getting it wrong, is always among people in concerted sequences of action who know how to take up the instructions discourse provides and to find, recognize, and affirm, or sometimes fail to find, what discourse tells is there, as well as relying on just such dialogic sequences to settle disputes about what is . . . Knowledge is always in time, always in action among people, and always potentiates a world in common as, once again, known in common.

From the perspective of sociology, then, concepts such as the dialogic account of knowledge, thought styles and thought collectives, and social construction open up for us the possibility of telling the truth about what we find through our systematic inquiries.

Dangerous Icons: From Magic and Religion to Science and Law

If magic is the extension of religious-based systems of explanation and manipulation of the physical world, then science is a secularization of magic, and the law or legal system is a secularization of the moral orders once also the purview of religion. "The Law" is a codification of social rules governing conduct and is a binding and hegemonic institution similar to science and facing parallel conflicts. For example, our project in this book is to present an anti-foundational or anti-essentialist approach to understanding technoscience. The anti-foundational projects in law, which draw from literary studies (Baron and Epstein 1998), critique the internal consistency and literalism of legal representations. This critical legal studies movement is of course greatly troubling to constitutional literalists, who are sure that all of our problems would be solved by keeping to the exact interpretation of the "founding fathers" of the constitution. That, of course, assumes that the founding fathers themselves were in complete agreement as to the meaning of the constitution. They were not, which is why the Bill of Rights was created as a set of Amendments and Federalists and Anti-Federalists had such a long-standing and bitter debate. More than two hundred years later, some lawyers and judges hope to discern the real meanings of the constitution and apply it consistently, in a wildly different historical context. It should be obvious that we cannot. Similarly, hermeneutics (the close study of texts for consistency and referentiality) challenges biblical literalism, and constructivist science studies challenges claims to literal "readings" of nature.

On the one hand, it is fairly easy to recognize the social construction process of specific laws. They have legislative histories and leave extensive paper trails in legal documents and sometimes the popular media. Some laws often have traceable legacies to other cultures or the more distant past. But "The Law" as an institution often seems outside human construction, and we sometimes respond to it as if were outside of ourselves. Some argue for "natural law," responding to scientific research as if it were an unambiguous referent for human conduct. Others look to scriptural sources as deity-given references for laws. A thorough sociological understanding of the law and legal system recognizes that as a social institution, the law works because there is basic social consensus as to its validity and legitimacy. While people may not agree with a specific law or its current interpretation, the system holds as long as problems appear accidental or as temporary errors, rather than systemic injustices directed at broad categories of people.

Science, as a term, is an abstraction for organized inquiry, and is a legitimating cultural idea for a hegemonic institution similar to law. The legitimacy of this icon functions in the same way that the cultural configurations of the west have become binding institutions, the yardsticks against which other cultural formations are measured. As a hegemonic institution, the institution gets to define the terms of debate about itself. While we might recognize that *science* is in fact sciences, that biology is not the same as physics, and that definitions of science can be manipulated, it is still clear that for many people *science* is an iconic term that stands for all that is rational and good in western society. Given these associations, then, to be critical of science is to be perceived of as being against the rational and the good.

Technology is particularly emblematic for US mainstream culture. Innovation and invention are part of the American cultural heritage, appearing in strategic places in the Constitution: the United States is an invented nation, rather than one emerging slowly through a long cultural evolution. Americans think of themselves and their technology differently from other people and their material culture in that we define ourselves by our access to it (English-Lueck 2002), and it is thus very difficult to think critically about technologies. Technoscience has an iconic quality for many that also maintains its parallel career to religion. Critics of religions are denounced as heretics and outsiders, critics of science and technology as dangerous cranks.

To recognize that truth is socially constructed does not mean that we don't know things, but that we can recognize, reflexively, that the things we know, while useful and reliable, are first of all contingent: they are contingent on the continued stability of social and physical systems. For example, even the simplest laws of physics rely on assumptions about the uniformity of matter and its properties, and on assumptions such as frictionless surfaces or ideal gas particles. More concretely, to *know* that antibiotics cure bacterial infections means relying on the continued availability of penicillin. Penicillin is an institutional achievement and its continued efficacy becomes questionable as bacteria gain resistance to drugs. That is the foundation of the second part of our realization: that truth is a social achievement and can be contested. Harry Collins (1985) has discussed the "experimenter's regress," where the validity of an experiment can always be called into question, leading to future tests, whose validity can be questioned, and so on. Consensus is achieved not because there is simply more evidence, but because an entire framework becomes accepted that fits into a larger worldview. Those who will not abide by the consensus can be silenced or moved to the margins.

When we begin to understand that "the law" is socially constructed, or the Constitution is subject to interpretation, or technoscience does not produce absolute certainty, or there are different interpretations for sacred texts, it does not make them useless but more responsive to society and its members. Each succeeding generation must take on the interpretive and meaning-making work and continue the practices that ensure the utility of ideas and systems. When this understanding is lost, people generally become afraid and defensive, rather than questioning those social institutions. In the case of science, many of our most basic concepts about knowledge need to be defined in sociological terms, not, as noted above for religious thought, in individualist terms. So, for example, while it is certainly desirable for individuals to try to be objective, objectivity is a property of a social group that comes to consensus about the identity and properties of a thing. Similarly, truth, and good and evil, become terms that are based in reference groups which define the terms and the criteria on which they are ascertained.

The simplest part of the worldview we are trying to explain in this chapter is a basic sense of humility: that the ability of human groups to understand and manipulate nature is powerful – and flawed. We need complex mechanisms to increase the reliability and validity of knowledge claims, while weeding out claims that are self-serving, dangerous, or otherwise wrong for the community. Understanding science and technology as social institutions allows us to understand the history of technology in a different way and also to understand present technoscientific activities and look to the future with different tools. Given our goal of thinking critically about science and technology, asking how it works, and for whom, we then have the opportunity to reinvent technoscience. So, for example, reinstitutionalizing technoscience asks us to look for new social roles or to understand old social roles differently. AIDS activists, for example, became very expert in scientific knowledge as well as about science policy and of course about the impact that HIV has on people's lives. In the process, they evolved a new social role. These activists blurred the boundary between "lay" and "expert" and helped change the way that drug research is conducted in HIV clinical trials (Epstein 1996). Reinstitutionalizing science and technology may also mean getting different people involved in doing technoscience. This can be a general call for diversity, to have more people from different walks of life contributing to science as it is currently practiced. It can also mean, however, looking at knowledge production in different venues, from the standpoints of people's ordinary lives, and taking those experiences seriously rather than dismissing them as "subjective," "anecdotal," or too simplistic. For example, instead of dismissing Lois Gibbs as "just a housewife" as she gathered information about health and illness and drainage patterns in the case of Love Canal, public health officials could have taken her research seriously (Savan 1988). That may have led to faster identification of the illness patterns, less exposure and fewer new illnesses, and quicker resolution of the case.

Conclusion

When we interact with institutions, particularly the state, we become part of the process of interpellation. This complex term describes the processes through which institutions label us as subjects, putting us into various social positions that then describe what we can do, and what can be done to us. These can be categories like race, gender, or class. Imagine going through life without ever having to check in a box for "M" or "F" to designate a sex, or another box to indicate your "race." Have you ever been convicted of a felony? Used drugs? These are examples of how subjects are constituted through interpellation, whether or not we wish to use these categories to identify ourselves or accept the constraints and expectations of the labels. Other distinctions range from credentials as experts or degree-holders, to class positions based on income and education. "Male" has long been an unmarked category in technoscience; until very recently, it was simply assumed that all scientists were male and one didn't need to think about what difference that might make. Science as an institution helps produce interpellations of various sorts, labeling people by gender, race, genes, sexual orientation, blood type, disease- or risk-status, and so on. Once labeled, our credibility, our scope of action, and our identity can then be systematically engaged. Obviously this process makes us, the authors, a bit nervous about the current world/technoscience we are creating.

The perspectives we are advocating here increase the awareness of professionals working within technoscience by helping them to understand the social relations and consequences of their work. It will no longer do to put off discussion of difficult issues, or make optimistic announcements without real research, or defer to other experts concerning the impacts of new technoscientific opportunities. The responsibilities of both the general public and technoscientific professionals will be intensified (cf. Bijker 2001). Of course, in the spirit of critical thinking, this approach requires that we make new demands of people who call themselves "experts," and question the notion of expertise more generally. So, when someone says, "trust me, I'm an expert," we are all now required to say "says who?" and ask questions about their training, evidence, assumptions, funding, and general point of view.

More generally, we have been concerned with establishing grounds for understanding what we mean when we say something is *true*. Truths are community property. They are therefore rooted in histories, social relations, and culture. They are constructed out of our interactions with each other in cultural and environmental arenas. We have sought to reveal the social nature of truth by exploring the way it is manufactured in magic, science, and religion. By raising questions about truths, it has not been our intention to challenge the very idea of truthfulness. The global intellectual movements that have made some thinkers claim that we now know too much to believe that is possible to tell the truth ("after postmodernism," some would say) have in fact taught us how to tell the truth (Smith 1999). The disagreements about this extend beyond the scope of this textbook, but the reader should know that this split exists and know where the authors stand. Everything we believe, everything we know, everything we do comes from our earthbound social and cultural experiences. The histories of sociology and anthropology can be read (explicitly beginning in the works of Nietzsche, Durkheim, Weber, and Marx) as an unfolding rejection of transcendental and supernatural explanations and orientations. We don't deny other readings or historical trajectories, but this is our reading and our trajectory. Marx claimed that all criticism begins with the criticism of religion. This chapter has been motivated by the idea that without a critique and a theory of religion we can't fully critique and theorize science. The reason is that any barriers to inquiry into beliefs, knowledge, and truth anywhere in society skew and undermine our efforts to understand and explain beliefs, knowledge, and truth in general. There may be an analogy from within the sciences. Mathematics was traditionally the arbiter of the limits of the sociology of knowledge. As long as barriers to social studies of mathematics were in place, we could not fully exploit our tools and perspectives for understanding scientific knowledge, especially the kind of knowledge labeled "pure."

The barriers to inquiry that have been set up around magic, science, and religion generally have served the interests of the more powerful in societies. This has led to contradictions within progressive social thought, inasmuch as scientific thought provides some of the ammunition for opposing religious institutions and political hierarchies. Critiques of science itself are seen as destabilizing those challenges (Croissant and Restivo 1995). Yet, as science itself is constituted as an institution, it suffers from the same pathologies and incoherencies of all complex institutions. This inhibits the application of the tools of science to science itself (Wright 1992).

Further Reading

Law, J. (2002). *Aircraft Stories: Decentering the Object in Technoscience.* Durham, Duke University Press.

Nader, L. (ed.) (1996). *Naked Science: Anthropological Inquiry into Boundaries, Power, and Knowledge.* New York, Routledge.

Nelkin, D. (1995). *Selling Science: How the Press Covers Science and Technology.* New York, W. H. Freeman and Co.

Philip, K. (2004). *Civilizing Natures: Race, Resources, and Modernity in Colonial South India.* New Brunswick, Rutgers.

4

STS and Power in the Postmodern World

Technology and Society

In times of rapid social change, social commentators try to explain and guide change, or challenge the assumptions and directions of change. Writing in Manchester, England during the middle part of the nineteenth century, social critic and theorist Karl Marx (1964, 1970, 1974) focused attention on the social inequities he saw emerging during the rapid industrialization and urbanization of the first "industrial revolution." Some one hundred years lager, Lewis Mumford (1934, 1964), Ivan Illich (1973), and Jacques Ellul (1964) wrote about social change in the twentieth century, dominated by continuing urbanization, mass production, and other key features of the modern era. More recently, Langdon Winner (1977, 1986) has written about contemporary relationships between technology and society from the perspective of political philosophy. Reviewing the ideas of these five social theorists and critics will help us explain some of the ways in which we can analyze, evaluate, and criticize the relationship between technology and society. Our decision to focus on these particular writers is driven by their influence, the relative clarity of their visions, and the lucidity of their writings. We follow this discussion with a look at the intersection of techno-social change and race, gender, and class. Our objective here is to illustrate some key ways in which critical research in STS has revealed and assessed the connections between technology and social life .We are not at this point endorsing any of these approaches, nor

73

is it our goal to give a detailed and thorough representation of each theorist-critic's ideas.

In *The German Ideology* (1947, 7), Marx and Engels, reacting to the abstract idealism of the Young Hegelians, establish their analysis on the foundation of a set of what they describe as "real premises." The first of these premises is the very "existence of living human individuals." It is the physical organization of living individuals who we understand as the system of cooperating individuals, that is, society. The production of means of subsistence arises out of this physical organization of human beings. In the process of working in and on the material world as social beings, humans begin to differentiate themselves from other animals. This differentiation can be considered the beginning of a process that leads to the distinction between society and nature, a distinction that is no longer obvious to modern students.

Technology and society are intimately, intricately, and reciprocally connected in Marx's analysis. If we understand, for the moment (we will amplify this idea below), technology as the set or system of tools humans use to organize for production, then technology is a critical factor in the self-identification of a people as a society. Furthermore, it is only societies that can create technology. Once this mutually dependent relationship between technology and society coalesces, the technology–society nexus begins to evolve and specialize. "Definite individuals" begin to produce in a particular manner and to create "definite social and political relations." In *The German Ideology* and other writings (e.g., Marx 1844/1958), Marx and Engels argue that social organization, technology, and the material conditions of social life determine human thought and the products of human thought. They do not, however, empower human thought itself with causal creative efficacy. The relationship between technology and society becomes more complex as the mode of production becomes more specialized and promotes a division of labor.

We need to go a little further here to establish that the idea of "technology and society" is too simple for Marx. Our claim that he views technology and society as reciprocally implicated means that technology does not stand alone for Marx; it is, in fact, a factor in the means of production, part of what gives rise to social structures, institutions, ideas, and values. So it is the relationship between the mode of production and the social relations of production that is the driving force behind societal development and social change. As the division of labor evolves and pre-capitalist and then capitalist social formations come onto the historical stage, the forces of production appear increasingly to be independent of the individuals who are in fact their creators. This happens as the division of labor splits up and opposes individuals, one from the other, even as the forces of production continue to provide the bases for the intercourse and association of humans. Here we have the seeds of, indeed the definition of, primary alienation.

As the division of labor progressively alienates the worker from the means of production, the process appears to remove agency from individuals and agency seems to accrue to the productive forces including technology. To simplify the matter, technology now increasingly seems to control production, individuals, and social change. Technology in this broadly simple sense comes to be seen as a power external to humans and to society. As technology is abstracted from the production process and takes on the attributes of agency (causal power), individuals become abstracted from their communities and society as commodities in a caricatured system of economics. The interests of individuals begin to lose their connections to the interests of society. In *Capital* Marx (1867) offers what is arguably the most detailed analysis of this process of alienation and commodification that makes humans appendages to the machinery of capitalism. This is a problem because Marx's image of the human being is broadly humanistic and, in his writings, he constantly detailed the errors and horrors of capitalism while running off hints here and there about a better, more humane society. If the division of labor has enslaved humans, the solution is to dismantle the division of labor. In fact, this meant dismantling capitalism and reunifying the interests of individuals and communities by bringing the primitive communism of early human societies into a modern version undergirded by the advanced technologies of the industrial revolution, advanced communism. Individuals integrated into their communities should be in charge of machines and not pawns of technologies under the direction of ruling classes.

Lewis Mumford (1934) began to write about technology and society during the period between the two world wars. He began by noting that advanced machines were in use in European society for seven centuries prior to the industrial revolution and that other cultures had the tools and machinery of modern industry. It thus appeared that other cultures besides the west had the preconditions for an industrial revolution. He explained the occurrence of the industrial revolution in the west by arguing for the compatibility between western values and the attributes and characteristics of the machine. The question he raises is whether the "inner accommodation" the west makes to the machine was in fact a surrender. Thirty years later, Mumford (1964) claimed that capitalism had over-exploited technology, documented the rise of the megamachine, and lamented the failure of society to integrate humanistic values and morals with its technologies.

In his earlier book, Mumford (1934) describes the machine as a product of human ingenuity and effort. Ideology and technics were fashioned together by a regimented society engaged in mechanizing life. The machine became the basis of a new religion in a society dominated increasingly by a loss of faith and purpose. The discipline of the machine reflected a desire on the part of the more powerful segments of society to establish power over others. This "holy" objective, not technical efficiency, was the motivation behind the development of machines. Whatever the objectives underlying the development of machines, one of the consequences of this development was to give some people the means to dominate other people and nature. Machines, according to Mumford, contributed to the need for order and predictability. All cultures are concerned with the problem of establishing and sustaining order, but the interesting thing about Western Europeans is that they sought order, regularity and certainty in technology. Technology and the discipline of the machine are readily experienced as external to their creators. As external forms and forces, technologies and machines appear to be outside the realm of morals and values. Mumford wants to see the mechanical order and the larger order of our lives integrated. Technologies should be subordinated to our values and our lives and not used to subvert or otherwise undermine our social worlds. Keep in mind that one of the issues this type of analysis raises is the possibility that the technologies we create can somehow be out of synch with our values. We should be asking whether and how this is possible.

Mumford's claim is that pre-industrial society sought order internally, thereby enabling individuals to control their technologies. The industrial revolution shifted the relationship between people and their technologies, externalized machine discipline, and gave tools, technologies, and machines control over their human creators. This created a situation that could readily be experienced as a shift in agency from people to technology and lead to the idea of technology as an autonomous agential force. Again, if we keep in mind that technologies are social relations (social institutions in David Dickson's [1974] terms), it is always going to be a matter of some people controlling others based on control of material resources. Technologies can be tools of social control, but they are not control agents. Nonetheless, it is important to understand that people have this feeling about tools, technologies, and machines.

Jacques Ellul (1964) struggled with the same sorts of problems that troubled Mumford. Ellul, however, focused his attention on collective mechanisms and movements in society that disenfranchise the individual in a very broad sense. Ellul claims that "primitive" humans were socially determined whereas modern humans are creatures of technological civilization. The driving force of a technological civilization is technique, and it is technique that integrates machine and society. Technique at once creates a technological world and helps adapt humans to that world. Once technique enters into all areas of life, it becomes the very substance of the human being, something inherent in all of us. There is some kinship here with the Marxian idea of a primitive communism in which technology and humanity are within, at one, with each other. While Marx, however, would see this as a function of the structure of human labor (social structure and material conditions of life), Ellul views the relationship between technology and society as a function of our ability to reason merged with technique, the set of means available in a society, This merger is the basis for efficiency, which Ellul values but defines narrowly.

Technique is ubiquitous in human history, but its presence waxes and wanes in societies. Among the Romans, for example, technique was precise and tied to achieving social coherence. In the sixteenth century and later, technique was restricted to the mechanical spheres and missing in other areas of social life. In earlier societies, technique was concrete and embedded in craft but later became more abstract, mathematical, and embedded in industry. The industrial revolution aimed to systematize, unify, and clarify all aspects of life and work. Ellul had trouble explaining this revolution because he thought of technique as internal to humans and society. Industrial society and modern technology thus took on a new and unfamiliar look out of touch with nature, artificial, and apparently autonomous.

This autonomous world seems to privilege technological agency over human agency. But the significance of the attribute of reason is that it implies agency. Ellul is not clear about this. Human agency is apparently a feature of "primitive" societies, but something happens historically – what and why are not clear to Ellul – that allows technological agency to gain control over human agency. In any case, Ellul's faith in religion provides a means for him to separate humans from "la technique."

Ivan Illich (1973) adopts an approach to studying technology and society that begins by imagining the attributes of a "convivial society" (a society in which tools and technologies are responsibly limited). Conviviality as an ethos refers to individual freedom and personal independence. Illich imaginatively explores whether and how tools are beneficial or destructive societal means. In this way, he is able to philosophically consider the limits and scales of the relationship between technology and society.

The creation of a convivial society depends on the possibility of identifying limits and relationships of scale. Industrial societies are built on the logic of large-scale production and engineered social relationships. In Illich's view, technologies per se cannot be good or evil; these are attributes of technologies-in-use. He makes a distinction between tools that help create a demand for what they can do and complementary enabling tools tied to self-actualization. One assumption here is that everyone has equal access to all tools and technologies. Illich believes, therefore, that it is possible to create machines that do not make us feel controlled or enslaved. Furthermore, this would support the sense of human agency over the mechanical world. The call for convivial tools is then a call for new tools, tools to work with rather than tools that "work" for us.

All four of these authors struggle with the idea of technological determinism, the idea that technology drives social change and that technological change is inevitable. So, on the one hand, they recognize that new technologies come from inventive activities informed by social needs or interests, needs or interests that reflect the views and social position of specific social groups. On the other hand, they all recognize the subjective feeling we all have to one extent or another, that we are slaves to the machines around us. In the face of new technologies, we seem to be faced with only two choices – take it or leave it. Some of us may even feel that technologies are forced on us at work in our daily lives. The political scientist and philosopher Langdon Winner has addressed this problem directly.

Winner (1977) takes on the belief and the feeling that technologies are autonomous; that is, they follow their own paths independently of our wishes and desires. Can technologies really be "out of control?" Winner argues that in fact technologies are not autonomous. How is it that we humans create technologies and then seem to lose control over our creations? To understand this apparent dilemma, Winner claims we must give up the assumption that inventors know their inventions in a precise and detailed way and can control their creations. In addition, we must give up the idea of a neutral technology. Technologies foster certain interests and goals while closing off and even destroying others. Winner (1986) has brought into focus issues of social responsibility in technological invention, design, production, and use, and the connections between technological systems and the human environment. Technological societies are complex, and it is in great part that complexity that is reflected in the transfer, transformation, and separation of technologies from human needs and creative intelligence. We need to get our bearings and stop sleepwalking in the process of technological decision-making. Once again, the answer to "the problem of technology" seems to be a call for integration. We recognize that there is already integration on some level, but it is problematic because of contradictions and inconsistencies in the way our society is organized not because technology and society are somehow independent entities that can go their own ways.

Unification and integration are the watchwords of social critics, such as Marx, Mumford, and Winner. They favor what we could call "secular" political means to achieve these objectives, whereas Ellul pins his hopes on religion. Illich, who falls into the "secular" camp, draws our attention to the difference between "working with" and "working for," a distinction that can be paired with the concepts of "power with" and "power over" (see the discussion below). The theme of power is at the core of all of these inquiries, whether it appears in the rhetoric of agency or of influence, control, and responsibility.

Before we look at the issue of power in greater detail, it is worth remembering that the issues we have been discussing are among the perennial concerns of social theorists, critics, and philosophers. Consider, for example, the ancient myth that has Zeus giving Prometheus and Epimethius the task of empowering mortals. Epimethius assigns physical powers to all the animals but forgets humans. Prometheus, however, famously gives humans fire and technical skills (stolen from Hephaestus and Athena). Zeus has to eventually send Hermes to give humans "justice" because the godly arts alone cannot insure that people will live peacefully and cooperatively together. Already in their early history, humans are struggling with the "wildness" of technology. The ambivalence about technology that this suggests (recall that Zeus repaid Prometheus for his kindness to humans by having him shackled to a mountain side where each day an eagle would work to devour his liver, each night the wounds would heal, and each dawn the eagle would begin again and so on forever) is reinforced by Plato's separation of science (*episteme*) and technology (*techne*), a separation that privileged science over technology.

Power issues are subtly present in all of the writings mentioned, but to understand the profound inter-relatedness of technology and society, an in-depth look at the type of power inherent in technology and society needs to be explored. Winner suggests the incorporation of values into technology, and recognition of the politics and values already designed into technologies which help give them their appearance of autonomy. Humanistic values are missing from technology as all of these men have demonstrated. Other values based upon control, order and domination of nature are the defining features of our technological society (Horkheimer and Adorno 1993).

The discussion about convivial tools and disillusionment with autonomous technology creates a space for questions about the relationship between technology and society, and questions about labor and identity in humans and machines. In the wake of the sorts of inquiries we have reviewed, social scientists have raised the following questions about identity, roles, and labor: Is the performance of labor, or a task, simply captured by its mechanical aspects? Do humans bring something else to their work? What is the identity or essence of humankind? Of machines? What purpose does labor serve for humans beyond addressing problems of subsistence? Asking what it is we want from our machines can seem to be a radical idea, given prevailing beliefs about technology at large in the wider culture.

Technological determinism is the belief that technology is the driving force behind other changes. For example, the statement "Computers will result in students losing mathematical abilities" reflects technological determinism. Somehow, the computer is the cause: not student behavior and attitudes, or teacher expectations, or the way that the computer is implemented, just the social fact of the computer's existence.

The media is full of examples of the pervasive myth that technological change is inevitable. Most popular media, business journals, and even films represent this myth. However, if technological change is inevitable, it is very difficult to explain the millions of patents in the Patent Office that are not being used: these are, for the most part, *technically* feasible technologies whose bells will ring or whistles tweet or gears turn. But the patents have never been taken advantage of and the technology is not available. There are complex factors that must be considered for a technology to be something other than fantasy or curiosity and actually be part of people's social worlds. For example, think about science fiction writing. It is fantasy. However, our fantasy of fifty years ago has either never been realized and remains fantasy or, in some instances, it has seamlessly become part of our everyday world as with Dick Tracy's wristradio or the gadgets found in the Hammacher Schlemmer catalogs.

First of all, a new technology must be manufacturable to begin the process of becoming part of people's worlds. It does not matter if there is an invention that can solve some sort of pressing social problem if it is prohibitively difficult or expensive to manufacture. Persistent problems with delivering annual influenza vaccines illustrate this issue, leading to lags and lack of coverage for different regions and populations. Commercialization is the process of taking a technology from a prototype or model and developing a way of integrating it with existing manufacturing infrastructures or developing new manufacturing systems. Related to manufacturing infrastructure is use infrastructure. In the US, we have 110–120 volts at 60 hertz alternating current at our domestic electrical outlets. It is of course different in other countries, leading to adapters for consumer products, and important design modifications for other technologies. Technological standards, such as internet protocols, can have that kind of limiting and enabling effect on a new invention. Without standards, there would be a proliferation of incompatible systems, but once standards are adopted, some kinds of innovations are no longer feasible, even if they have advantages over existing systems. At their origins in the 1960s, early microwave ovens needed to be installed as part of a kitchen's cabinetry and electrical system in a new home or by expensive remodeling, rather than being the rather portable appliances of today. If people cannot plug it in effectively, it is unlikely to appear on the social landscape as much more than a curiosity or rare luxury item.

One aspect of technological change not appreciated until recently is the idea of cultural convergence. If a new technology significantly disrupts important cultural ideals, it is likely to be rejected by potential users. For example, the drug RU-486, used for non-surgical abortions, is also useful for treating conditions such as endometriosis and may be useful in breast cancer treatments. Its adoption in the US and approval by the FDA was significantly delayed because it conflicted with anti-abortion activists' values, despite its utility for other medical conditions.

Cultural convergence issues can also explain why technologies that make sense in western contexts fail when applied in other cultures, leading to rejection or disruption. The technology "works" in an instrumental sense, but fails in a cultural sense, leading to non-adoption. For example, Japan has a sophisticated medical system very much like the western allopathic system of formalized, science-based medicine. However, organ transplantation is exceedingly rare, and undesirable in Japan unlike in the United States and parts of Europe. Different ideas about when death occurs and about the bodily integrity of the person (Lock 1996) mean that cultural and religious values inhibit acceptance of organ transplantation as a viable medical treatment in Japan. The Japanese do not generally subdivide death into a series as westerners do in distinguishing between brain death and organ death. One is either dead, or not dead. Brain death is not recognized as some sort of intermediate step, and to remove organs from a person with no measurable brain function is both disrespectful to the person so afflicted, and perhaps even akin to murder. In this case, the mere existence of a technology (organ transplantation) does not determine if, and how, it will be incorporated into a society.

Consider the ambiguities of the "Green Revolution" of the 1960s and 1970s as Western agricultural interests exported industrial agriculture to the non-western world. In many cases, in the short term, yields and nutrition improved. But these beneficial effects were short lived as the costs of the new systems – fuel, new seed stocks, fertilizers, and pesticides – became unsustainable. Steve Lansing (1991) studied Balinese traditional water-sharing and crop rotation systems integrated into the social life and religious calendars, which were disrupted by "modern agriculture." Very quickly pest problems got worse. Fields, which had been burned or flooded periodically, were kept in production and harbored pests; conflicts over water and water shortages grew as sharing mechanisms (both social and material–technical) worked out over centuries were disassembled to promote continuous resource use. Research in other farming communities has shown that people would prefer low but reliable yields to high but fragile or easily disrupted production. People now struggle to reclaim traditional, sustainable agricultural practices, and this may give some indication as to the hesitations over the so-called "next Green revolution" in genetically modified organisms. "Green Revolutions" also have a tendency to seek single robust seeds and high yields, a tendency that leads to monocultures vulnerable to pests and diseases; diversified ecologies are not as easily wiped out by a single pest or disease.

And, of course, a technology must *work* to be adopted. This leads to the questions of "works for whom" and "what is meant by 'works'?" The obvious answer is that a technology works when people turn it on and it does what they want. But what they want can be social or symbolic as well as practical and functional. An expensive car, such as an SUV or luxury model, can be fuel inefficient, unsafe, and polluting but still provide a social signal that the owner is affluent and sophisticated. We can argue on a socio-cultural level that cars do not "work" for our society, with pollution, urban sprawl, and accidents eroding the quality of life, but few can imagine alternatives at this point in time. Or let us consider small kitchen appliances: the true function of a kitchen gadget is its symbolic value as a gift. Such gadgets make nice wedding presents, but they are not necessarily that useful to the people who receive them: they sit in their boxes until the next yard sale. These issues lead to problems of technological reconstruction. While designers and marketers may think that they have a good idea about how a new technology will fare, users in different social locations may have very different ideas about what a technology means, how they can or will use it, or even if they will use it at all. Advocates of the social construction of technology (SCOT) argue that the "working" of a technology is only partly explained by its technical functioning: the fact that bells ring, gears turn, and whistles whistle does not explain a technology's successful proliferation, but must be explained by social factors or integrated with social explanations. So, given this kind of complexity, it should be very difficult to argue that technological changes are inevitable. Change is highly contingent. The pervasiveness of technological determinism as a myth leads people to be compliant in the face of possible changes, rather than questioning them and perhaps influencing how technological systems are designed and implemented. It is interesting to look at who perpetuates the myth of technological determinism and to see what they have to gain by sustaining the myth. It is common knowledge that "the minute you buy a computer it's outdated." What makes it outdated is (1) our desire, belief, and values that inform our view that the next one will be "better," and (2) an innovation system that is based on the relationship between hardware and software manufacturers, not consumer needs (Rochlin 1997).

The idea of the "technological fix" – the idea that more technology, or a piece of hardware, will solve a social problem – is an important element of technological determinism. The idea we want to stress is that social problems have material elements (for example, resources such as land or money), and that technologies that are inserted into social relations have within them assumptions about social relations and causality. Consider the idea of moving into a gated community or building a wall around your own house as a technosocial solution to a perceived problem of personal safety. One of the elements in this is privacy: people cannot see what is within the walls, so, it is less likely they will be tempted by things that do not belong to them. However, law enforcement officials know that the privacy afforded by a wall also allows things to happen behind it with less chance of a passerby intervening. When burglars go over that wall, they can take their time because they are *also* protected by the privacy of the wall. The wall does not address the root problems of criminality (complex to say the least) and decreases the kind of community safety and identity that helps to make places safer. The wall is an inadequate "technological fix" that misdiagnoses a social problem.

Rather than technological fixes, or technological revolutions, what we often see is a process of technological intensification: a social process is intensified as people select technologies which allow them to reach their goals. For example, analysts have argued that the car and the telephone were complex consequences of the rapid urbanization of and very rapidly changing geographic mobility in the United States at the turn of the twentieth century. However, major waves of urbanization and migration predated the availability of either technology. It is clear, however, that as people moved across the country and into cities, automobiles and telephones alleviated the negative consequences of mobility, allowing people to stay in touch with distant family and friends, for example, or avoid social isolation and dependence on public transportation (Fischer 1992). One could argue that cell phones, as annoying as they are in public spaces, are (with Walkmen and comparable portable devices) technological "solutions" to public isolation and a breakdown in ties of social solidarity as communities fragment. They also intensify social processes that support or facilitate instant gratification and connection on demand. However, modern technologies like cell phones, Walkmen, and PDAs (personal data assistants) have simply made more visible and vivid cultural landscapes that emerged long before these devices.

Another pervasive myth surrounding technology is that it is morally neutral. The same technology that goes into a commercial bakery oven can also be implemented in an oven for destroying people in a concentration camp. That does not, however, make the technology neutral. It does, however, indicate the possibility of technological reconstruction by users and also demonstrates technological flexibility. Madeline Akrich (1992) has argued that technologies come with scripts for their use, developed explicitly by designers and often modeled on analogous tasks. However, users rewrite or disobey the scripts, often as sources of innovation or to solve local problems with implementing the technology. Sometimes, this creates problems. Technologies are designed, which means that they embody the interests and values of their builders. Sometimes, these are not very well thought through or articulated, and so misuse is not anticipated as a technology moves through different social worlds. An example of this is the technological reconstruction of a commercial aircraft into a projectile explosive. Airplanes are designed to carry volatile fuel, be steered in particular directions, and move quickly. Other things, like guided missiles, are better designed for those functions, but their access is more tightly controlled, and there are not many degrees of freedom in their flexibility. The same affordances of an airplane that overlap with a missile can be used in ways not anticipated by airplane designers, but they, nonetheless, reflect values and preferences: speed, control, flight duration, and explosive potential.

A common phrase used to argue for technological neutrality is that "guns don't kill people, people kill people." If technology were truly neutral you should be able to substitute any other technology for the word "gun." Try: "Sofas don't kill people, people kill people." This somewhat silly example highlights the fact that we know sofas are not designed to kill people, and guns are. One could kill someone with a sofa, but its affordances are such that it would be a very difficult task. And of course people indirectly kill themselves by spending too much time on the sofa at the expense of their health. The concept of affordances is thus a term for the things that are designed into a technology to facilitate its use for certain activities. Guns are "meant" to be fired. Their handles and parts are organized and shaped to make it easy to expel a high-speed projectile at another body or object, a projectile that cannot be recalled. That projectile has been fine tuned for values such as tumbling to maximize damage or accuracy and penetration. You could use the stock or butt of a firearm to pound in nails to hang a picture, but you might damage the weapon (and the wall, nail, and picture). Affordances can be overridden or ignored, but that requires special efforts to overcome the designed "intentions" of the system, the basic values reflected in the technological design, and prevailing cultural meanings attached to the technology: it means rewriting the scripts. Thus, we need to get beyond pronouncing a technology "good" or "bad" or "neutral" – but look at the intentions of its designers, its affordances, meanings, and the social relations of its production and use and its short and long-term impacts to ascertain whether the technology is worth keeping around. To say that context is everything might sound like we are excusing the technology, considering it as neutral, but technology is part of the context and shapes the interactions of participants in a social situation when actors interpret the meaning and affordances of a technology and act with it.

Actor-network theory speaks to the ways in which technology constitutes the context of interaction, by positioning material objects, whether scallops or firearms, into kinds of actors with features and characteristics which shape their use in the context of human actors trying to achieve their objectives.

Power, Values, and Agency

Technology studies focus on cultural products and mainstream society and acknowledges the non-neutrality of technology for minority groups (Wajcman 1991). By examining technology's participation in the meta-discourse, a deeper understanding of power, agency, and values is possible.

The works of Marx, Ellul, Mumford, Winner, and Illich show how technology is created by humans living in societies. The type, manner, and application of the technology varies within and across cultures (Pacey 1976, 1998; Basalla 1988). One classic example is the stirrup that facilitated war on horseback in medieval society where the "Anglo-Saxons used the stirrup, but did not comprehend it" (Finn 1964, 24) while their rivals did use it to gain superiority in warfare. The same invention in China (like many inventions that originated in China) had no transformative impacts on Chinese culture comparable to their effects on the West. The power of the emperor and the relative homogeneity of Chinese culture and ecology absorbed much of the potential impacts of new technologies (Karp and Restivo 1974). Post-World War II Japan's recovery was fueled by a unique style of industrial organization, management, modes of problem formulation, and attention to detail that enabled their chemical, electrical, and machine-tool industries to expand and surpass the same industries in the west (Pacey 1998). Thus the "same" technology imported into a new culture will be modified to work within the new context.

Other scholars in other fields have also articulated how people and their material productions interact in the realms of art and identity. For example, the reproduction of art through mechanical means has made it more accessible to the masses, who now absorb art instead of art absorbing humans (Benjamin 1968). Art in paintings, photography, or film is a cultural product that has adopted and incorporated technology. The politics and ideology of art have been manifested in different forms at different times and in different places. Technologies are implicated in politicizing art because technologies are social relations. The nature of actors and audiences and their relationships is integral with the technologies of art and social change. In live theater there is a certain intimacy in the interaction between the audience and actor, but in the age of mechanically reproduced theater or movies that specific form of intimacy is lost (Benjamin 1968).

The computer's entrance into modern life has also raised new questions about the relationships between people and machines. Sherry Turkle (1984, 1995) traces how people use the computer as a way of creating an identity for themselves. Her questions center around how individuals "stand in the world of artifact." Clocks and mechanical transport have changed societal notions of time and distance. The artifact named "computer" is changing our conceptions of mind and self (Turkle 1984). Turkle argues that the changing of mind and self by computers is not determined by, but is evoked by, computers. Her work focuses intimately on questions about the nexus between identity and machine.

> We search for a link between who we are and what we have made, between who we are and what we might create, between who we are and what, through our intimacy with our own creations, we might become. (Turkle 1984, 12)

Turkle highlights the creation of identity for various individuals through the computer, bringing into the present Marx's insights on how we use our manipulation of technology to produce our selves.

One aspect of identity that she does not pursue very far is gender relations. Turkle (1984) acknowledges that there is a gender divide. The majority of the people she interviewed are male. People who generally like to master things, have power over something, and gain pleasure from this manipulation typically interact well with the computer. While many men and women work *with a computer* at work and at home, men are still heavily involved in working *on*, i.e., designing and building computers. This distinction between *with* and *on* is important. Control, power, and flexibility are more evident to those who, besides using computers (or any technology), can manipulate and change them (Croissant 2000). Even while software development, management information systems, and computer science show signs of increasing gender-equity, electrical engineering and hardware design are still largely the domain of men. Turkle claims that it is "a flight from relationship with people to relationship with the machine – a defensive maneuver more common to men than to women" (Turkle 1984, 210). What are the full implications of this: that people who like to master something are socially awkward and that it is generally males who utilize computers for mastery? What does that say about our society and the growing use of technology? What does this mean for the portion of society that is excluded from this technological world?

In her work on identity, Turkle (1984) explores the idea that the relationship between technology and individuals can help capture the full essence of humankind. Arnold Pacey (1983) hints at the idea that it is technology that strives to capture the essence of human nature:

> One idea behind all these inventions was the dream that if one could make a clock or other instrument that exactly reproduced the motions of the sun and the planets, one would capture something of their essence. (36)

It is interesting to think about the possibility that humans mimic nature by creating technology to fully understand who and what they are. This is exemplified in cultural studies of technology. For example, literature is another tool we can use to explore the relationship between culture and technology (cf. Leo Marx 1964; Winner 1977). Leo Marx identifies a cultural space where literature, art, myth, and narrative converge. He explores the idea of the "pastoral" in literature in order to demonstrate changing representations of technology, nature and society. These representations develop and change as societies and cultures evolve. Ideology,

Marx shows, is an important determinant of a society's perspective on technology (Marx 1964). Similarly, Winner (1977) uses literature throughout his book to illustrate ideas of autonomous technology. His discussion of Mary Shelley's novel *Frankenstein* highlights the need of "those who build and maintain the technological order . . . to reconsider their work" (Winner 1977, 316). He also compares it to the Hollywood version and analyzes the cultural implications of the differences. Rosalind Williams (1990) examines the idea of the "underground" environment, one which must be highly technological for people to survive in, or the result of a technological disaster as humans are driven underground. The idea of the underground in literature shows us, among other things, how our apparent dependencies on technology make us nervous. The literary scholars, and many others who do cultural studies of technology, demonstrate what Wiebe Bijker (2001) calls a "semiotics" of technology: a focus on meanings ascribed to and arising from the use of technologies in practice that take on weight and stability over time and influence the way that technologies are used and new technologies are imagined.

Questions about how to integrate technology and society tend to revolve around images of and ideas about the self.

> Doing, making, producing, – technology in its general sense – are activities that clearly help shape our sense of self . . . If we give technology its fuller meaning of knowledge and process, we find it, first, filling our time and governing our movements. (Cockburn and Ormrod 1993, 159)

The first relationship highlighted by Cockburn and Ormrod is the one between the individual and technology. This is a relationship of empowerment. The individual is actively producing something with technology that contributes to his/her identity. There is power for the individual in the technology. This is reminiscent of Marx's description of technology and society as internal to one another. A second relationship implied in the quote is that of society being governed by the technology. In this relationship there is a sense that technology has power over or dominates the society.

This parallels Marx and Mumford's description of technology as something that is experienced as external to society, something outside ourselves. This shift also impacts our perception of the locus of agency. It now appears that technologies and communities possess agency over individuals. This is a feature of the shift in scale from small locally controlled artifacts to large-scale systems tied up with national, regional, and global systems of political economy. These changes once again bring questions about agency to light. What is the role of agency in the relationships between technology and society? How do we want to define our relationship(s) to technology? Can we design technologies that are more environmental and people friendly? How do we deal with the conflicting values that drive societal interests in constructing technologies? And is the concept of agency still viable in the light of our knowledge about social structure, society, and culture. On the one hand, discussion about agency can be more usefully framed as matters of freedom. The concepts of agency and freedom tend to quickly tie themselves into philosophical and logical knots in the face of structural and material conditions. Freedom, on the other hand, has a substantive significance in politics, the economy, and social life. Do we want to say: I have agency and can write whatever I like; or do we want to say: I am not forced by clerical or secular agents to write what they want me to write – I am not imprisoned. I do not write with agents of the state or the church looking over my shoulder? Is it a question of having or not having agency over technology, or (as we believe), a question of having the political and economic freedom to influence design, to not go along with a new technology imported in your office by your boss, to resist the real people behind imposing, advertising, or selling a particular technology?

The problem of power in the technology–society nexus is not an easy one to master. Power is an issue in the relationship between individuals and institutions as well as in the relationship between technologies and individuals. It might be helpful to think about two different forms of power: "power over" and "power with" (Kreisberg 1992). An individual doing, making and producing is engaged in power with ("co-agency"). If we experience technology as reshaping our activities, thoughts, and meanings, then we feel as if technology has power over us; we feel dominated by and disciplined by machines (cf. Foucault 1979). We are trying to open lines of inquiry with these ideas and questions. However, the reader should keep in mind that analyses of technology and society can be carried out without using the concept of agency and with more sophisticated conceptions of power that do not invest things or people with agency. Flows of behavior within social and cultural boundaries can be understood without invoking agency by treating power as a matter of access to and control over pools of resources.

Ruth Cowan (1989) discusses how the use of tools places limitations on the work people do, and how they do it. Much in the same way that Marx noted that people can become appendages to industrial machines, even in housework, the affordances of tools often shape the way we organize our work. For example, the adoption of the family car altered the locomotion of choice for family members from walking, bicycling, and public transportation, resulting in one adult (typically the mother) becoming the family chauffeur (Cowan 1989). When tools control the worker, or when technology has power over an individual, then the tool could be said to have technological agency. Those who control tool-making generally have power over those who use the tools. In many technological societies, it is elite men who are tool-makers, and through these tools, they are able to control the labor of women and other subordinates. A factory is designed by architects, engineers, and upper-management to fulfill the goals and objectives of the company, while it is manual laborers and operators who run the machines on a daily basis within the factory. The tool the worker uses (which is already a set of social relations) determines how, where, what, and when s/he places her/his body in space and time. Tools are also implicated in who has access to what machines for what purposes. The idea that the tool has power over the worker is widely acknowledged as is the power of design to determine the type of labor. Designers and artifacts are often invested with agency. However, if we learn to see tools, machines, and technologies as social relations, then it becomes easier to see people with varying degrees of power and freedom behind the workings of technology and society.

Cyborgs, Humans, and Technology

A fully technosocial theory would take into account the reciprocal influences of society on technology and technology on society. The historical validity of technosocial theory is already clearly established by Marx, Mumford, Ellul, Illich and Winner, and elaborated by insights on the social construction of technology. But what is next? There are two major streams of thought in this area, both trying to avoid dualisms between nature and culture, and between humans and the things they make and use. One is cyborg theory and its variations. Donna Haraway (1991), in her "Cyborg Manifesto" argued that we are always already "cyborgs" – cybernetic organisms that integrate mechanical artifacts and biological human action into our identities and human societies. Whether one wears glasses or contact lenses, or uses a prosthetic or a wheel chair, or experiences a tool or a car as an extension of the self, people are everywhere manifesting the concept of the contemporary cyborg. The cybernetics of the self is linked to another feature of contemporary society, fragmented identities. We live in overlapping social worlds with multiple social roles and identities, some of which conflict. For example, we face many choices as consumers for creating an identity in terms of the things we use to define ourselves. In sociological terms, this is the difference between achieved and ascribed characteristics, the former being ones one works for, the second identities one is born to. These have been in flux in the twentieth century, in what is called the postmodern era. One of the issues is the fluidity and flexibility of identity, and the kind of individual and public anxieties that this might produce. Computer communication, with avatars and chat rooms and anonymity, exemplifies and facilitates this: one never knows exactly whom one is communicating with, and one can adopt an alias or explore other identities freely online.

While the cyborg blurs the boundaries between humans and machines, actor-network theory (ANT) does the same, perhaps with a less inventive imagery. ANT models technosocial life as a web or network of actors, humans and artifacts, which impinge on one another, and facilitate certain kinds of interactions. Technological systems emerge from the relationships among different scales of users: individuals, work groups, businesses and business sectors, and the state. Artifacts are kinds of actors, because their properties act upon and constrain the desires and possibilities of other actors such as humans, who are simultaneously acting upon and constraining the properties of artifacts.

But all of the most recent models of technology, whether cyborg theory, the social construction of technology, or actor-network theory, are incomplete when it comes to understanding power among human social groups. For example, a simple social determinist would argue that technologies merely reflect the existing power relations: a society based on residential segregation invents gated communities to prevent the "wrong" types of people from being in a neighborhood (a form of argument found in Winner's work). A more technologically determinist argument would be that "the internet" makes less-powerful people more equal to people with more power, because it is readily available and not currently overtly controlled. A more nuanced view would recognize that technological development often follows the lines of power: money, influence, expertise, force. Technologies can also, of course, be used and subverted by the less powerful. Cyborg imagery seems liberatory, because cyborgs can refuse to be either human or machine, and can choose their identities and components. Cyborgs in film, fiction, and reality, however, are almost always owned by someone else: they are property that only gives the illusion of complete choice and rights as they strive to become fully "human." Even Pinocchio wanted to be a "real boy."

Cyborgs, then, are ambiguous figures of power, and power and property are two of the major variables which shape technological decision making. Take the problem of risk and the distribution of the "goods" and "bads" of intensely technological systems. It is fairly clear that toxic by-products of technological development are most often located near economically marginal communities, often with high concentrations of ethnic minorities. This can be partially explained by economics: the land is cheap, so people choose to bear the risks of living near hazardous sites in order to own homes. But often people who have to put their wastes somewhere select these poor communities because, economically speaking, the land is less expensive, but also because the poor are less connected to media resources and political power, are less well organized, and have less income with which to protest or prevent such sitings. Thus, the people who can least afford the costs of increased health risks are the ones most likely to face them. All sorts of other systems have similar kinds of effects, which challenge ideas about uniform progress through technological change.

Risk analysis is becoming an issue for technology studies because of the increase in high-risk technologies. Assessment of risk is carried out from individual to industrial levels to raise questions about accountability, predictability, acceptable levels of risk, and policy implications. Underlying any assessment is always the question of "how safe is safe enough?" and who decides (Morone and Woodhouse 1986). It is not just expert communities, but lay people, users, and workers who need to be asking these tough questions, seeking alternatives, developing research strategies, and learning from previous mistakes to successfully analyze risk.

Charles Perrow's (1984) *Normal Accidents* addresses risk assessment at the industrial level and points out the impossibility of control on organizational and technological levels. He argues that complex systems are fundamentally untestable and unknowable. Complex systems are often tightly coupled, meaning interactions in one part of the system can lead to unexpected events in another part of the system. Safety devices and monitors can themselves become sources of problems. Organizationally, there arises a tension between extreme routinization and training to reduce operator error, and allowing for flexibility and decision-making power by operators so that they can react to problems not anticipated. Perrow suggests that perfect technical and social control are illusory, and thus some complex systems may be too hot to handle.

Increasing self-knowledge and fuller awareness were thought to come along with increasing technical control . . . Then suddenly technology itself came under attack as the source of danger . . . It became plain that the old link from danger to morals was not made by lack of knowledge . . . The difference is not in the quality of knowledge but in the kind of community that we want to make, or rather, the community we are able to make, or I should say, the community that technology makes possible for us. (Douglas 1992, 9)

Douglas asks us to consider a risk analysis that acknowledges the human factor both as individuals and as it is worked out through the power and structures of institutions. She mentions these two aspects of the "human factor" that can modify and affect the way research is done. Winner (1986) suggests, however, that the language of risk already cedes much to the definitions used by the powerful. In using a cultural theory lens, Douglas (1992) is able to achieve insights into the "control of knowledge, the emergence of consensus and the development of expectations."

The emergence of AIDS in the late twentieth century forced many who may have perceived their lives as safe to reevaluate the risks in daily life. Media images and peoples' perceptions of AIDS and the immune system have been studied to understand common sense, risk, and identities (Martin 1994). After interviewing a diverse segment of Baltimore society, Martin (1994) found the metaphors being used by individuals to describe their bodies and lives was based on a model of the self/world as a complex system. Systems thinking is a new worldview to our society, one that is based upon our experience with technology. One of the effects of systems thinking, like functionalism in mid-century sociology, is to prioritize the stability and harmony of the system. This makes the group central, and individuals bear the brunt of adapting to the system, rather than the other way around. People are to be ever-flexible and willing and able to adapt. This is an emblem of our postmodern world. For example, consider how "hot" chaos theory is as a popular method for explaining technological problems. Systems theory emerged starting with World War II research into cybernetics, information, and control theories, and also helps to shape our postmodern world with its tools and models. Thus, this theory is useful as a critical metaphor for understanding our technological civilization, at least as long as the human elements of power are remembered.

Contemporary Society: Globalization or Bust?

The one thing that individual nations still command is its citizens and their intellectual prowess. Science was once an activity of citizens who happened to live in a state, although they were often social elites whose identities were not necessarily strongly tied to national identities. For example, Kepler practiced his science wherever he could find an environment that encouraged his inquiries. In today's global society, science is an activity of the state. Kepler would not be as free to move to a more liberal country like the Denmark of his time if he lived in the twenty-first century. States control the movements of individuals, information, and technologies. This shift in science and technology from an activity of networked individuals in scientific cultures to nations and states managing science requires a new form of analysis.

In the twenty-first century we are experiencing a highly interconnected world achieved through the removal of boundaries that have traditionally differentiated nations and people and the travel and articulation of things: products, technologies, corporations, and industries. In the "new" global economy, nations are sharing services, values, and products defined by the technological and scientific cultures. No longer are goods and services produced, utilized, sold, and disposed of locally. For this "global economy" to work, standardization must be introduced to facilitate exchanges across "traditional" borders. "Worldwide activity is nothing more than high volume standardized production transplanted abroad" (Reich 1992, 121). We must recognize that alongside this movement that seems to homogenize cultures, local cultures and states can become more clearly delineated so that some observers have introduced the word "glocal" (Robertson 2001) to express the complexities of the modern era. George Ritzer (2004) has extensively, and critically, documented this transformation of consumer products, while Drori et al. (2003) have looked specifically at these transformations as scientific institutions and models migrate.

This is also an era variously described as the "information society" or the "knowledge society." There are debates about whether there is indeed something informationally distinct about our era. We have not thought it appropriate to devote much time to this issue at this introductory level. However, readers should be alert to this perspective on the global society (see Webster 2003; and for an advanced critique of the information society idea, see Lash 2002).

In the late eighteenth through the early twentieth centuries Europeans were engaged in expansion, colonialism, and imperialism. Their navigational technology in instrumentation and ships enabled them to travel to new places that were exotic, different, and fascinating. One way that the difference was evaluated was through comparing scientific and technological achievements of their home country with those of the countries they visited. When scientific knowledge and technological applications were measured, evaluated, and compared by the Europeans, they found the science and technology of savages to be lacking and a basis for declaring the unknown culture inferior. This defined the measure of worth of a society as its level of science and technology. "Machines were the most reliable measure of humankind" (Adas 1989, 134). In true scientific fashion, criteria were established to measure the scientific and technological knowledge of a culture. This was "the most meaningful gauge by which non-Western societies might be evaluated, classified and ranked" (Adas 1989, 144).

"Modern" science and technology have been used as a lever of power over nature and other peoples. westerners were able to assert domination over African and South American peoples because of their "superiority" in weapons and naval technology. However, this is not a "necessary" outcome of science and technology. Gunpowder and explosive technology was present in China three centuries before it appeared in Europe. Iron casting was mastered in China fifteen centuries before it was mastered in Europe. In 1420 the Ming Navy was the largest navy in Asia with trading ships, galleons, and warships. It would have been a match for any European state or alliance. However, due to a change in state policy, long distance navigation lost support, and by the sixteenth century the navy was only a ghost of its former self (Needham and Temple 1986). The Chinese did not apply their science and technology in the same manner as the Europeans. Different cultural values, politics, and resources helped to shape the way European science was applied and used. Science and technology were a source of power for the European explorers; it was often a mystery, a marvel for the "natives" they encountered. African and South American cultures were fascinated and awed with various mechanical items of European travelers. This mystery was used to dominate them. They were considered inferior because they lacked the western scientific worldview.

The Europeans judged the Africans to be "guided by instinct rather than reason" and that their abstract reasoning was thought to be limited (Adas 1989, 118). Therefore, it was the responsibility of Western people "to bring them into accord with modes of thought and behavior the colonizers deemed rational, efficient, and thus civilized" (Adas 1989, 205). In other words, the colonizers attempted to make them into little models of themselves, to eliminate differences, and to convert them to Christianity, a rational religion. "European colonization [was] the triumph of science and reason over the forces of superstition and ignorance which they [Europeans] perceived to be rampant in the non-industrialized world" (Adas 1989, 204). The Europeans used their science and technology to "civilize" the "uncivilized" and it worked because it capitalized on the beliefs and superstitions in these cultures, sometimes by demonstrated technical efficiency and often by force.

The communities that joined together for freedom and liberation created a new identity that embraced all the members and united them within their differences. The nation becomes a nation-state when it is bureaucratized. As unity became uniformity, the nation-state identity shifted again to a national security state. Science is an integral part of a nation becoming a nation-state and national security state. As boundaries and categories become more concrete in a nation, the maintenance of rules brings violence to the forefront of the nation-state's activities and responsibilities: warfare and welfare. Somewhere in and in between warfare and welfare are science and technology projects. Shiv Visvanathan (1997) tells the story of India, a modern nation-state and how science, technology and development projects begin with good intentions and end with the loss of civil liberties. He highlights the absence of victims from the history and discourse of science on various technological projects in India from Bhopal to dams and nuclear parks.

Visvanathan (1997) ties together science, violence, and the state through an exploration of the "banality of [the] everyday." He labels development as a scientific project based on the scientific method, belief in progress, a "vivisectional mandate" (where the Other is the object of violence in the name of science), and "triage" that marries rational experiments, obsolescence and vivisection to label and judge local knowledges as obsolete and incurable. He critiques modernity as an escapee from antiquity that has yet to face the Other. The violence of modernity is not only state violence but "the violence of science seeking to impose its order on society." The violence of modernity can be seen from a wide variety of examples from plagues in India and Egypt to Nazi science. Visvanathan argues that social triage is being applied to development to the benefit of some and the disadvantage of many. In social triage the Other is identified as dispensable, dangerous, and threatening to the center. These Others are also the ones least likely to have a voice in modernity or technocratic projects. He argues that underneath the ideas of the modern state and science is an intolerance for people or knowledge. This intolerance originates in the fear of someone successfully challenging the power of science or the modern state. Visvanathan (1997, 281) cites various dam projects around the world to show "development as slow genocide" where "peaceful development has created more refugees than have bloody wars." He concludes that development based on modern science is an anti-ecological force or a form of terrorism.

Visvanathan (1997) names the seed as one of the most important metaphors of modern society as it is memory, past, present and future. In opposition to the seed, there is science which has no memory because it is constantly subjected to updates and revision. If science had memory like a seed, he suggests that science might embrace diversity, rather than attempting to eliminate it. Visvanathan laments the knowledge that is lost in the classificatory activities of science that serve to make more room for science itself. The justification for forgetting or erasing knowledge is that it is "un-scientific." This also marginalizes any scientists who may attempt to seek out the diversity of Other knowledges. The discourse of development uses the language of healing where Western nations are the healers and "third world" nations are the patients. This discourse of healing constructs and legitimizes development as a scientific reality and enterprise. The role of modern medicine is to ignore and discredit folk medicine and its infrastructures as it did with midwives and herbalists in Europe in the nineteenth century.

Metaphors, Narratives, and Glocal Cultures

Visvanathan (1997), along with other scholars, advocates paying attention to local and expert knowledges and their uses of metaphor, narrative, and authority in developing a critical understanding of science, power, and culture. Local knowledge is opposed to "universal" or "global" knowledge. "Local" implies pluralities and relativities. An ecological locality can be a village, a region, a country, or a global sector. The locus of local knowledge is the practical problem in space and time socially contextualized. Linking sets of practical problems can produce a problem area and spawn specific technologies (e.g., alternative technologies). To the extent that local knowledge spans problem areas distinct from those addressed through "expert knowledge," narratives may become useful ways of trading knowledge.

"Narrative" has become one of the tools and metaphors of science studies and cultural studies of science. Postmodernists tend to apply the term to any piece of meaningful discourse. In this context it is used as a way of exploring and explaining science and technology – as narrative per se, as another form of fiction, and as practical discourse. In the context of local–global discourse, a narrative attempts to make sense of some local event or emergent. Such narratives must be temporally situated and report on emergence in terms of origins or histories. One of the assumptions underlying this approach to narrative is that there are distinctions between the human, biological, and physical sciences in terms of "emergence." This could, for example, be reflected in the harmless-looking distinction between the human "domain" and the biological and physical "sciences." This reflects a prejudice, not a reasoned classification. Viewed over a sufficiently long time span, all areas of discourse will appear at one time or another as domains rather than sciences, as domains at some points and sciences at other points, and at some times controversially perched between domain and science. The very process of inquiry, subjected to anthropological inspection, may in fact not sustain this distinction from moment to moment, day to day, or fact to fact. This implies that emergence and novelty are not restricted to occurrences across domains and sciences. They may, indeed, be punctuated on a time line at different rates of appearance along different segments.

The point of science studies is that scientists' accounts of their own work cannot stand on their own but that they must be measured against other ways of constructing accounts of how science is made. The sociologist of scientific knowledge doesn't simply produce another alternative account of science, a narrative of science opposed to science unfolding "naturally" and without the assistance of metaphors and narratives. Indeed, the sociologist in the laboratory becomes a contingency and in the end constitutive of the institutionalization of reflexive social theory into the very fabric of what science is and what the scientist is. Eventually, the disjuncture between sociologist of science and scientist disappears, and they become as one in authoring the invention and discovery of facts about the world. It is possible to see this phenomenon of the sociologist-in-the-laboratory as just another one of the hybrids, boundary transgressors, and classification confusers characteristic of the contemporary world. It is one source of questions about how local productions get "universalized" or "travel," and how global interests find their way into local settings. This sets the sociologist-in-the-laboratory firmly in the midst of discussions about the general relationships between the local and the global.

The local–global disjuncture is not entirely new as a focus on intellectual concern. It appeared during the middle part of the last century in social science discourse as the micro–macro problem, and in an earlier era as the gemeinschaft–gesellschaft distinction. It entered anthropological discourse following World War II as a focus on people, communities, and states in an increasingly international world. During the 1960s, it was emblazoned in the slogans about local and global, people and politics. Now as we began to turn our attention directly to this old concern in the context of thinking about the nature of knowledge, science, and technology, we came across an essay by Benjamin Barber on "terrorism and the new democratic realism." Barber (2002, 12) writes that "What alone has become clear [in the wake of September 11] is that we can no longer assign culpability in the neat nineteenth-century terms of domestic and foreign." Nonstate actors, whether they are multinational corporations or loosely knit terrorist cells, are neither domestic nor foreign, neither national nor international, neither sovereign entities nor international organizations.

In one way or another, our own theoretical concerns about local and global in knowledge and science studies must reflect the revolutions and transitions of twentieth century global political economy. There is a resonance between Barber's view and those of other contemporary political theorists who write about the local-global disjuncture (see below), and science studies scholars who want us to focus not on "local" knowledge but on "located" (situated) knowledge (e.g., Biagioli 1996).

The twentieth century witnessed an increase in the scale of human activity and awareness that enveloped the world. This increase in scale is the root of a variety of issues, problems, and questions captured in the concerns across the human sciences with the local–global nexus. Regional increases in socio-cultural scale have occurred throughout history. In the fifteenth century BCE, for example, a cosmopolitan civilization emerged in the Middle East as changes in scale broke down geographical and cultural barriers (McNeill 1965; cf. Sjoberg 1960). Such developments have often been accompanied by ideas about world unity (Wagar 1967). The linguistic roots of the one world concept can be traced to the Cynic word, *kosmopolis*; the practical roots of the idea lie in actual or dreamed of increases in socio-cultural scale. Some version of this idea is generally part of the ideological toolkit of conquerors and is grounded in an expansionist, imperialistic, and dominating orientation to peoples and cultures outside the conquerors' current sphere(s) of influence and control. Alexander, for example, conceived the goal of his conquests to be the establishment of *homonoia*, that is, of human concord among the nations he conquered.

The idea of world order also occurs in the writings of philosophers and is generally grounded in an idealistic view of the basic unity of all human beings. For example, Zeno of Citium, the Stoic disciple of the Cynic Crates of Thebes, conceived of a world ruled by one divine and universal law. One needs to be careful when distinguishing the "one world" ideas of military adventurers and those of philosophers and theologians. The latter are, after all, responsible for providing ideological and mythological defenses for conquering heroes and their armies. Somewhere out of these rearrangements of territory and cultures, states and peoples, systems of knowledge and science emerge and change.

The distinction between local and global may be fading with the fading of the political categories that Barber identifies. The local may have to be understood as somehow interdependent with the global; globalization may always involve localization. Thus, we are given the neologism, "glocalization" (Beck 2000, 45–47; and see Robertson 1992, 1994). Hardt and Negri (2000) have even suggested a way to link the local to the universal:

> The concept of the local . . . need not be defined by isolation and purity. In fact, if one breaks down the walls that surround the local (and thereby separates the concept from race, religion, ethnicity, nation, and people), one can link it directly to the universal. The concrete universal is what allows the multitude to pass from place to place and make its place its own. (263)

Conclusion: Technoscience and Globalizations

Large-scale systems, often organized and controlled by the state, are the most obvious places where power and control are concentrated, and where the idea of technological determinism rings the most true. But even then, we can dig behind the finished facade, the stable infrastructure, to see the values and compromises built into our dams, power grids, internets, or buildings. There we see how power was congealed momentarily into a new building or sewer line or highway. The durability and scale of these systems should not be taken as inevitable, or proof that power is a one-way operation. To the extent that we agree to go along with these systems, to forget that we have some responsibility for their continuation and maintenance, or to forget to apply our imaginations to thinking about how things might be different, then we cede our power, however small, to these larger institutions.

One of the important characteristics of the twentieth century was glocalization of knowledge and science. If this process – borne on the shoulders of imperialism, colonialism, world wars, and on the expanding tentacles of corporate and media networks – reinforced the concept of a universal science, it also revealed the extent to which the universal is a creation, a construction of the everyday/everynight world of soldiers and priests, CEOs and university professors, investment bankers and television executives, actors, actresses, poets, athletes, tourists, and scientists themselves. As the technosciences escape their local confines to be embraced by or imposed on the Other in the Elsewhere, they do not do so in simple, single, unilinear, and unidirectional narratives. The invention of science is already the reinvention of science – and this reinvention of invention repeats itself in every era. And as these processes unfold, increasingly embodied in and carried by information, we humans flow into new identities, new roles, and new places in the world and in the universe.

Everywhere, the local is shaped by the global. In the laboratory, people, resources, and symbols flow in and out along network tracks that reach into every corner of the world. Laboratories are crucibles within which the new world social order's image of life, its new image of the technosciences based on a networking logic (cf. Castells 1998, III, 345– 378), and its new creation myth are being fashioned. In the next chapter, we discuss the construction of socially intelligent machines as a mode of reproductive technology. The control and distribution of reproductive knowledges and practices are "contested in every society" (Ginsburg and Rapp 1995, 5). Social robotics research, developments, and applications are already spreading across the information networks of the world. This globalizes the conflicts over modes of reproduction. Transformations in new reproductive technologies define the locus of intense cultural antagonisms. Information flow thus becomes an important vector for moving the technosciences around the world in a multilinear, multicultural dance of dialectical fireworks. The new narrative begins: "In the beginning was INFORMATION . . ."

Further Reading

Casper, M. J. (ed.) (2003). *Synthetic Planet: Chemical Politics and the Hazards of Modern Life*. New York and London, Routledge.

Rappert, B. (2003). *Non-Lethal Weapons as Legitimizing Forces? Technology, Politics, and the Management of Conflict*. London and Portland, Frank Cass.

Sinclair, B. (ed.) (2004). *Technology and the African–American Experience: Needs and Opportunities for Study*. Cambridge, MIT Press.

Terry, J. and Calvert, M. (eds.) (1997). *Processed Lives: Gender and Technology in Everyday Life*. New York, Routledge.

Turnbull, D. (2000). *Masons, Tricksters, & Cartographers: Makers of Knowledge and Space*. London, Tayor & Francis.

5

Life after Science and Technology Studies

What we have seen unfold in this book is a view of science and a view of technology as social constructions. Our goal has been to tell this story in a way that does not lead to relativism but does not, on the other hand, leave us in the wonderland of classical science worship. (Remember our example of the planet – or not – Pluto.) Things do exist, but only in terms of how we act on and use them. This applies to "reality itself." Even the most ardent critics of social construction are no longer defending the naïve view that there is a "thing in itself." We have emphasized that things exist within and surrounded by complex social webs of meaning, connotations, and denotations. Pluto exists as Pluto within a scientific culture that has instruments, scientists, and a particular cosmology. However, for a community outside of or on the fringes of scientific cultures, Pluto can only exist within their social webs of meaning. These social webs of meaning-in-use are located in and draw their significance from cultural and historical settings. To represent these complex socio-cultural webs of sciences and technologies, we have borrowed the word technoscience or technoscientific.

Insofar as science is a general term for politically unfettered thinking, inquiry, love of nature, curiosity and critical analytic approaches to understanding and explanation, we believe in science. But by politically unfettered we do not mean that science can be ultimately free of politics; it must always be understood as a political phenomenon, contextualized and constructed from micro- or interpersonal levels to macro-levels of the state and resource control. But we believe that inquiry is at its best when it proceeds without powerful people, without weapons, and without the power of the state standing in the wings or looking over one's shoulders, directing, controlling, and intimidating.

Can we children of the postmodern era ground ourselves in *physical* reality without becoming intoxicated by a Grand Narrative of Physics? Can we accept the reasonableness of looking both ways before we cross the street without succumbing to some reductionist and mechanistic model or theory of reality, and claims to political power based on false certainty? Can we understand that the moon is not made of green cheese without becoming seduced by Truth with a capital "T"? And then – can we ground ourselves in the pervasiveness and complexities of social relationships, social interactions, and social constructions without becoming intoxicated by and succumbing to a Grand Sociological Theory or a Grand Narrative of the Social? Consider the recent debates about the status of the planetary object called Pluto. Pluto, an ice ball at the edge of the solar system, does not fit into the two major models of planets as rocky balls or gaseous giants (Freeman 1998). It is smaller than some asteroids and other astronomical objects, and has a very irregular orbit. But Pluto has been embedded for nearly a century in professional definitions of planetness, and so defined in the textbooks and aphorisms of schoolchildren. Our claim is not that Pluto is not there, although its presence can only be detected through mediating instruments. It is that the "reality" of Pluto as a planet or planetoid, or asteroid, or bit of cosmic ice and dirt, is only intelligible within a framework and narrative of explanation, which is always already based in human interests. The "always and already" formulation hearkens to Althusser and his ideas about subjectivity (Althusser 1998) and reminds us that we can never get outside our own skins to claim some "god's-eye view" (what Haraway 1991, calls the "god-trick") of reality. Pluto does not and cannot care one way or the other what we decide about it, even if our decisions and frameworks lead to particular interventions. These are the challenges that face us as we travel into the twenty-first century and beyond, challenges that we will explore in the following set of case studies on new reproductive technologies and robots and society.

Case Study: The New Reproductive Technologies

Jaycee Buzzanca came into the world with six parents. There were the two anonymous donors of eggs and sperm, her gametic parents. There were her contracted parents, John and Luanne Buzzanca, who hired married surrogate mother Pamela Snell to carry this child. Snell's husband did not ever make a formal legal claim to being Jaycee's father, although he could have. Shortly before Jaycee's birth in 1995, John Buzzanca filed for divorce from Luanne, and refused to pay child support. An initial ruling by Orange County Superior Judge Robert Monarch supported John Buzzanca's argument that he was not the father of this child, since they had no biological relationship whatsoever, and on similar reasoning, neither was Luanne Buzzanca her mother, although she could adopt the child if she wanted. At one point, Snell also considered adopting the child she carried, but the initial ruling stated that Jaycee Buzzanca had no parents at all.

Shortly after the initial ruling, the California Court of Appeals for the Fourth Appellate District decreed that John Buzzanca and Luanne Buzzanca are the legal father and mother of Jaycee, three years old at the time of that ruling, and that John was liable for child support. Jaycee had lived with Luanne since birth. The basis for the court's conclusion was the rule that parental relationships *may* be established when intended parents initiate and consent to medical procedures, even when there is no genetic relationship between them and the child. That is, language in the California Uniform Parentage Act (Section 7610) does not exclude establishing parenting through means such as surrogacy and infertility treatments such as donations of gametes. The reasoning hinged on the use of the word *may*; traditional genetic relationships and formal adoption are not the *only* ways in which parentage could be established. The agreement to proceed with the surrogate pregnancy and use of anonymous donated gametes established the Buzzancas as the intended parents. It also clarified prior rulings, which argue that the surrogate mother (or egg or sperm donor) is not the parent of the child, the contracting couple is, because they were the initiators of the "procreative relationship."

Advocacy by various legal parties extended the courts' arguments to include the possibility of determining parentage before the birth of a surrogate child. That is, rather than having the contracting parents adopt the child after it is delivered, they can claim legal parentage prior to birth, and be listed as the parents on the birth certificate. Thus this step would save extensive delays and costs often accrued in "step-parent" adoptions, which is how surrogate relationships were often categorized. Surrogacy opponents, including anti-abortion advocates, generally agreed with the results of the final ruling. Those with a pro-life orientation decried, however, the use of anonymous genetic materials in the surrogacy, saying that it leaves no legacy for the child at all, and continues to treat reproduction as a contractual issue and children as commodities.

Other recent cases, however, continue to confuse the issue. In 1994 and 1995, for example, a New Jersey couple conceived one child via in vitro fertilization and stored the remaining seven embryos at a facility that promised to destroy the embryos if there was a divorce. The couple did divorce, and the biological father sued for possession of the embryos. As a strict Catholic who believes life begins at the moment of conception, he equated the destruction of an embryo with the end of a life and decided to take the embryos back, perhaps for future implantation in a new wife. His ex-wife fought his case, arguing her right not to have her biological children born without her consent. She won her case in the New Jersey appeals court. The right not to reproduce, despite other legal arguments that do not award parentage to the donors of reproductive materials, overrode other contractual language, especially since the court determined that the father in this case was capable of reproducing in the future. Other cases, however, have allowed women to implant preserved embryos, despite the protests of current or former spouses or their estates. But in other cases, the right not to reproduce has been upheld. For example, in 2000, a Massachusetts Supreme Judicial Court (Avila) ruled in favor of the rights of a father to prevent use of his genetic materials in frozen embryos. This upheld a Probate judge's determination based on the initial ruling following the divorce of a couple that had frozen embryos in storage, which the woman wanted to have implanted. In part, the contractual arrangements specified at the beginning of assisted reproduction processes help to determine the use, donation, or destruction of preserved genetic materials, but there is still a lack of consistency in the legal and cultural interpretations of these conflicts. These cases also became related to issues surrounding the use of embryonic stem cells and cloning, because, if the contracting and donating parties all agreed, unutilized embryos could be donated for scientific use.

These and many similar cases and issues need to be understood as more than a "cultural lag" problem, where the legal and cultural systems are somehow "behind" an autonomous science. And the individuals in these cases need to think of their dilemmas not as private troubles, but as public issues (Mills 1959). The "right" to reproduce, a cultural thread prevalent in Western and other cultures, helped to "pull" scientists into research on embryology, endocrinology, and reproductive medicine, to solve the perceived problem of childlessness for traditional families faced with infertility. As some women delay child-bearing to later in life when the chances for reproductive "success" decrease, new reproductive technologies (NRTs) seem to be a necessary "solution" to a new "problem." Techniques such as in vitro fertilization are often used with side-effects on women, when it is their male partner's limited fertility that is the "problem." To the extent that new technologies help to redefine families, allowing, for example, gay or lesbian couples to use their own genetic materials in reproduction, or single mothers to reproduce without the entanglements of a relationship, the demand for NRTs increases. As the NRTs have moved into other arenas and circumstances, the paradoxes, ironies, and deleterious consequences have become more apparent. So rather than thinking of the rest of society as "lagging" behind science, we should try to identify the cultural formations that push scientists into new directions.

We should look at NRTs as places where different institutions, namely science, the law, economic institutions, and families, come into contact in new ways. We can expect technologically enhanced reproduction to continue to produce controversy, and to produce new social roles and practices to manage the overlap. For example, there are now specialists in reproductive law, managing an interstitial space where contact between science and families is new and untested, and specializations in bioethics and medical ethics which discuss these new possibilities.

Many argue that these technologies cannot be banned, for legal and for practical reasons. The reproductive industry is highly unregulated, and so banning an NRT in one place will mean that it will pop up somewhere else, including in international markets. In addition, a rights orientation is used to argue that despite risks of premature births, risks to the mothers (who undergo drug regimens with known side-effects including increased cancer risk), the high financial and emotional costs and losses (the best clinics have a 65 percent *failure* rate for in vitro fertilization), we cannot ask people *not* to use NRTs if they are desperate enough. This of course begs an important question: why don't we challenge the cultural formation that makes them feel so desperate? Why, exactly, in a world with scarce resources, is not having a child a problem? Consider the alternative formulations: child-less, or child-free? Why is the language of possession, property, and ownership so central to the language of reproduction? For example, people want children "of their own." As if there were not enough children to go around, of various ages and circumstances, who need all sorts of nurturing.

There is also "ethical creep" going on: as new boundaries are breached, and new things become technologically possible, it becomes more difficult to argue against their use. For example, with pre-implantation screening of the genetic materials in a fertilized embryo it is now possible to increase the *probability* of defect-free births. There are still no guarantees! Sex-selection and identification of fetuses with known genetic disorders is more reliable, increasing pressures on pregnant women to abort undesirable fetuses either because they have measurable "defects" or because they are the "wrong" sex. Nonetheless, it is quite possible that individuals and societies will begin to pay attention to ethical creep and to turn away from technologies that are "bad for them." Perhaps we may yet learn to live without cars. Or we may, despite recent proliferation problems, decide that nuclear weapons development is no longer necessary. Harmful drugs may be pulled from the market. We do *not* need to commit to technologies that seriously undermine social or physical health. However, we must first be able and willing to start to ask "the right" questions about problems and issues we have identified.

Case Study: Robots, Minds, and Society

The development of socially intelligent and emotional robots has created a social space of border tensions between minds and bodies, machines and humans, and scientific and theological-religious authority. These tensions are not novel in the history of cultures, but the emergence of prototypical social and emotional robots has substantially elevated those tensions. We now have social and emotional robots such as MIT's Kismet and Yuppy, a bi-pedal humanoid robot (the Honda robot), common-sense savvy computers (e.g., Cyc), and computers that can already, according to some reports, pass the Turing test (Artificial Intelligence Enterprises' Hal). These developments come at a time when humans are becoming siliconized and machines are being carbonized. More and more body parts are being mechanically replicated and implanted in living humans, and organic substances are being used in the development of robots and other intelligent machines.

If we see human bodies as in some sense "special," and in particular if we see human bodies as in some sense *spiritually* unique (e.g., possessing a soul) then the tensions at the machine/body boundary will resist resolution. On the other hand, if we follow Nietzsche (1968) and claim that there is only body (that is, incorporate the body fully into the materialist paradigm), then we can think of bodies as interpretations. This conception of bodies as matters of interpretation seems to make a place for the possibility of new kinds of bodies with new kinds of inner lives, a place in fact that it is in our (will to) power to construct. The limits of AI and social robotics are not to be found in the limits of silicon and steel but in the limits of the interpretive courage or foolhardiness allowed by new social relationships. The anthropologist Levi-Strauss (1966) argued that academics focus their attention on phenomena just at the point that they are ending. However, what often appears to be endings are actually transitions and transformations. The contemporary focus on the body, then, is not a matter of the end of the body but rather the end of one kind or interpretation of body and the beginning of another kind or interpretation of body (Martin 1990).

The reason philosophical and logical traditions get in the way of social analyses is that they are grounded in and emerge out of research and theory in the sciences of the material and natural world. In addition, they reflect their roots in theological discourse. These discourses, universalized and generalized globally, consistently fail us in our efforts to understand ourselves and our social worlds. Randall Collins (1998) has stood out for decades as one of the great modern interpreters of social logic. It should not be surprising, then, that while he is, with Harry Collins (1990, 1998), one of the pioneers in transforming our understanding of the social nature of knowledge, science, and belief, he is also a pioneer in sociologizing the very idea of artificial intelligence.

The idea of a socially intelligent robot and of a robot that interacts with the world the way a human infant does (e.g., Cog and Kismet) arose not in a sociological laboratory but in the AI Lab at MIT (Breazeal 1999a, 1999b; Brooks 1999). Social theory, however, supports the directions of these efforts and in fact supports the very idea that machines can think, become conscious, and be emotional. If the claim can be made that some machines have already achieved such states, then we should be exploring how social theory and research might facilitate the further development of such machines to higher, more complex, and more sophisticated levels. The problem is that emotion, for example, is not definable in the isolated human or animal body (or "mind"), but that it is *only* defined within social groups. It is not located in a neural system or a bodily state. This is in fact the case for mentality in general (cf. Brothers 1997). Even "acts of perception" have a social context and are not unmediated "natural" acts. Euclidean visual space, for example, may be a Cartesian artifact, a consequence of the extent to which Western culture has "carpentered" its environment (Heelan 1983; Campbell 1964).

Our objective in supporting these efforts is not so much to create social and humanoid AIs and robots, but to further our understanding of human mentality as a social fact. In engaging social robotics physical scientists and engineers, we are concerned less with robotics problem solving than with issuing some challenges. Current research aimed at constructing social and humanoid robots will, so long as it is committed to psychologistic and cognitivist models and theories, only succeed in producing more and more interesting, more and more complicated, and more and more dangerous toys. The relevance of sociological theory to social robotics R&D is that it suggests that the most ambitious goals of social robotics researchers may be capable of being reached. This is not a prediction that we will have conscious, thinking, emotional machines by a certain date, or even ever. We do claim that the chances of succeeding in this field will be significantly improved if social robotics researchers adopt models and theories that are better and more broadly grounded in the social realities of mentality. Swarm models, while they are too primitive and contrived at this point to serve as a serious contributions to the social theory of mind, have in at least one case been grounded in a social psychology of mind (Kennedy and Eberhart 2001).

The collaboration between social psychologist James Kennedy and engineer Russell C. Eberhart (2001) on swarm intelligence represents a significant disciplinary boundary crossing in artificial intelligence research. Chapter 6 in their book *Swarm Intelligence* is dramatically titled (from the perspective of AI and social robotics research more generally) "Thinking is Social." From our perspective, it is significant that G. H. Mead (1934) is listed in the references (though he is not in the index and does not appear to be cited in the text), but not L. Vygotsky (1978, 1986). The authors explicitly understand that the mind is social, and this is more important at this stage of their research and our inquiries than their swarm model. It may be, as they claim, that swarms metaphorically model soft computing (computational intelligence) in a useful and insightful way. Kennedy and Eberhart themselves describe their work as at the "toddler" stage.

Robots – and especially social and emotional robots – are emblematic postmodern objects. They are loci and emblems of emergence, local knowledge, and narrative. Like new reproductive technologies, great expense has gone into solving sets of technosocial problems without asking whether or not these problems are well defined, and if perhaps other questions need to be asked. What drives the cultural fascination with social robotics? While the fascinating puzzles that robotics and artificial intelligences pose for scientists, engineers, philosophers, and sociologists are fairly obvious, the economic and social desirability of robotics remains elusive. For example, why is it that taking care of the aged is so often seen as a goal for social robotics? What are the social processes that contribute to the inability of some aging persons to care for themselves? What are the social formations that lead families to either struggle in seeming isolation with elder care, or for the elderly to become isolated from their families? Why, culturally, does no one seem to want to have older persons around, at least according to the social roboticists? What then, are the economic relations that make robots seem an attractive alternative to skilled and unskilled human beings as caretakers? What are the impacts on the many hundreds of thousands of workers who currently serve the elderly? Who will be responsible when robots fail or err in their care? Until these kinds of questions are answered, the robots produced will have limited application despite much hype and promise, and the root causes of high costs, disability, family stress, and alienation of the elderly will be unexamined and persist.

Frontiers and Horizons

In the early years of STS, researchers used to discuss the problem of the "hard case." In terms of classical sociology of science, which focused on the social system of science and left "scientific knowledge" out of the picture, the hard case for the new social studies of science was scientific knowledge itself. Within science, certain forms of knowledge were "harder cases" than others. In the early years of STS, mathematics was – along with logic – the hardest case, the arbiter of the limits of the sociology of knowledge. The idea that scientific knowledge was socially constructed quickly became so transparent–even irrelevant–in some STS circles that "social construction" was replaced by "construction" (cf. Latour and Woolgar 1979, 1986). Some of us continued to think, and to think to this day, that "social construction" was not as transparent as some people in the field assumed. Furthermore, efforts to eliminate "the social" by various postmodern theorists have been in our view premature, and to some extent just another form of resistance to social analysis. It has been our experience in the classroom, in the lecture halls, and at conferences that the idea of "the social" is not at all transparent, and that there are many sources of cultural resistance to the very idea (cf. Collins and Makowsky 1998).

We are still struggling with the problem of communicating the idea of the social in and out of STS proper. We have struggled here to communicate that idea while at the same time applying it in a difficult and complex context. The reader should be warned that there are some very good reasons to be suspicious of "the social" (of social analysis, of sociology), reasons which have been discussed by some of the most advanced thinkers of our age. It would not take much to convince us that we – especially we – would benefit from being more critical about the social. Nor would it take much to persuade us that new movements are afoot that are transforming the idea of the social as it has been handed down to us as heirs of classical social theory. Nonetheless, we stand by the continuing viability of the social as a mode of explanation and analysis, as a window into and onto who and what we are as human beings. In this age of hybrids and hyphens, we too must be prepared to transform ourselves and our "social" even as we defend our vitality as thinkers and the vitality of our ideas.

There was always something "dangerous" about the sociology of science, especially when it took on the task of explaining scientific knowledge. Sociology itself was a dangerous form of life and met much resistance long before it was applied to truths and facts themselves (Collins and Makowsky 1998; cf. Berger 1963). The application of sociology to the hardest case, science, and then to the hardest of the hard cases, mathematics and logic, made sociology more dangerous still. It raised the specters of relativism and of the Other. Slogans like "anything goes" in the methods of science, meant to sharpen our critical sensibilities in the face of totalizing discourses and Grand Narratives, became signs of threat to reason and truth and to Western civilization itself. This is part of "the rest of the story" and beyond the scope of this primer in STS. The next question we must ask ourselves is that having "solved" the hard case of science by demonstrating that it is a human performance, a social construction, have we become complacent and less dangerous? Indeed, there is a hard case beyond the hard case of science and dependent on developments in the sociology of science that is worthy of our attention. Religion could claim this place, but religion is no more a hard case any longer than science (Restivo 1991). A new hard case for science studies is "mind and brain." We spent some time in chapter 2 discussing the sociology of mind. Here we want to focus on the brain as a sociological problem.

While the 1980s were leading into the 1990s and the Decade of the Brain, various provocations were emerging that opened up the possibility of a sociology of the brain. The sociology of mind has a long and distinguished if invisible history (invisible in the schools and among the general public and much of the intellectual world), and we have rehearsed some of that history in earlier chapters. The recent history of the neurosciences is interesting for us because while it is focused on the sensually, materially present brain it is at the same time moving in directions that open pathways which sociologists may follow into the world of neurons and synapses. Even while the brain continues to reign as a powerful locus for explanations in the study of human behavior more and more brain research points to social and cultural influences.[1] Before we look at the sociological and neuroscience rationales for a sociology of the brain, let's begin by reminding ourselves about the sorts of brain headlines that regularly appear in our newspapers and magazines.

The major theme in the headlines we want to draw attention to is the strategy of "looking" into the brain to explain "mysteries" of the human condition. The major characteristic of this strategy is the confusion it reveals about just what it is we are talking about. The July 17, 1995 cover of *Time* magazine broadcast the headline "In Search of the Mind." The subheading read: "Scientists peer into the brain looking for that evanescent thing called consciousness." The cover illustration is a human head marked with squares and letters and numbers, and with a rectangular opening at the top of the head that reveals blue sky and clouds. The headlines suggest we are looking for mind and consciousness, but also that mind and consciousness might be the same thing; and what is "brain," then? Stories illustrated with open human heads are quite common in this genre. Newspaper headlines during the Decade of the Brain (the 1990s) heralded the brain at work: "Brain Yields New Clues on its Organization for Language" (Blakeslee 1991); "Photos Show Mind Recalling a Word" (Hilts 1991). Steven Pinker (1994, 1997) became a star of the mind wars with titles such as *The Language Instinct*, and *How the Mind Works*. Such titles hid some of the subtlety of Pinker's arguments, subtleties that were nonetheless too crudely biologistic to escape the criticisms of neurobiologists such as Steven Rose (Pinker and Rose 2004). Newspapers and magazines regularly published photos of the "mind" (why not the "brain"?)

recalling a word, and the search for the soul, God, and a moral center inside the brain. Many if not all brain research continues to be grounded in assumptions about localized functions, and these assumptions have been strongly reinforced by various brain scanning techniques that produce images of areas of the brain "lighting" up as subjects and patients carry out tasks. Thanks to sociologist of science Susan Leigh Star (1989), we know something about how localization ideas emerge and get sustained in the everyday work world of neuroscientists. Her work doesn't lay bare a history of falsehoods but rather a history of social worlds and work. Like much of the sociology of science and technology our book rests on, Star's work demonstrates the day-to-day and night-to-night efforts that sustain theories beyond the laboratory results themselves. Neuroscience, indeed, is an agonistic field of competing ideas and claims regarding localization. In the nineteenth century, phrenologists divided the surface of the brain into thirty-five regions, each impacting on our personality. Character could be determined by surveying bumps on one's head. It wasn't too long before phrenology lost favor among serious students of the brain. But some critics of contemporary localization theory call it neurophrenology. Without deciding the issue of what it is that the brain does, we can already see that neuroscience is ripe for sociological investigation. There is more grist for the mill of a sociology of the brain in the attention given in the United States to preserving and studying Albert Einstein's brain. It is widely assumed in science and among the general public that something interesting can be learned about creativity and genius by studying *Time* magazine's Man of the Century (although this is probably not as true today as when Einstein died). In the Soviet Union, the brain that has received the most attention as a source of a possible explanation for genius is the brain of Lenin (Abraham 2001). China has recently opened its first brain bank, and scientists plan to collect detailed psychiatric and psychometric information from potential brain donors. However, the program may encounter cultural resistance to the removal of brains from corpses as many Chinese prefer to bury the body intact.

The questions raised by the brain industry are these: first, why do discussions of brain often become discussions about mind, or consciousness? Why are mind and brain distinguished in some cases, but treated as synonymous in others? It is very easy to slip from one focus of analysis to another in this field, and that suggests that we ought to pay more attention to what the terms of our discourse mean, and what they refer to if anything; second, how can we sustain a perspective that makes the locus of explanations about human behavior the brain when it is obvious that without social life no behavior is possible? It may seem obvious that we are social beings, but the brain is often treated as an entity that behaves autonomously and as an entity "we" can interact with (suggesting that there is "me" and there's my "brain"). We are not free standing, independent, autonomous brains. Why not? The reason is that without bodies and without mobility there is no possibility of mentality. Only philosophers and science fiction writers could conceive of the possibility that we are brains in a vat and that all of our "experiences" are fed to our brains by aliens or computers. The brain's operation, whatever it does, is dependent on the social fact of our communicative interactions with each other and with our environment. The first major non-philosophical effort to link mind and brain to society was undertaken by a "brain scientist," not a social scientist. Psychiatrist Leslie Brothers (1997), internationally recognized for her work on primate social cognition, cross-connected the brain sciences and sociology in her book, *Friday's Footprint*, in ways that had not been done before. In laying out her objectives she writes:

> The classic fictional story Robinson Crusoe describes how a shipwrecked sailor survived alone on a desert island. He prayed, kept a diary, and industriously made tools, clothing, and shelter for himself. This image of the isolated individual embodies a metaphor for the human mind; it is the metaphor that has determined the practices of contemporary neuroscience until now. To bridge the worlds of brain and mind, we will replace this isolated mind metaphor with a view that is thoroughly social. (xi)

At this stage of her work, Brothers was still, for all of her sociological imagination, working in a brain-centered framework. So even though she realized that we needed to go "thoroughly social," she was still thinking in terms of brains "working jointly to make culture" (that is, she made the move from the isolated Crusoean brain to networks of brains). But it is networks of humans that we have to focus on. It is *culture* – human beings working and talking together – that makes brains, minds, consciousness, and thoughts.

One of the problems with theories of brain and mind, of mentality in general, is that we are all to one extent or another prisoners of mind/ brain, mind/body, brain/body dichotomies and dilemmas. We must get over the belief that when we meet Einstein's brain, we meet Einstein, as some people imagine (Paterniti 2000). We need a solution to these classical problems of mentality, and this may be the next hard case for sociology and science studies. We cannot pretend to be ready with a solution that would be at all satisfying to sociologists, neuroscientists, or anyone else interested in these problems. We do, however, think we can say something about the direction such a solution would have to take.

Restivo (2003) proposed a solution to the problem that took the following form. He claimed that we need to collapse the dichotomies and recognize the validity of Nietzsche's (1968) insight that there is only body. Restivo's strategy is to eliminate the possibility of recourse to transcendental entities that resist material reference, things like "soul" and "mind" and "consciousness." Such things are better understood as possessing symbolic reference, as products of certain ways of talking or as components of speech acts. Without getting too technical, the point is to pursue the idea in two stages. First, adopt the view that it is bodies that are the locus of mentality, bodies that think. Second, adopt the view that bodies are thoroughly social and that therefore when bodies think, it is communities that are thinking. It is not the grammatically illusory "I" who speaks, but society that speaks through you, thinks through you. We are beginning to see research results that provide direct and indirect evidence for Restivo's conjectures (cf. Brothers 1997, 2001; Valenstein 1998, and see endnote 1). For example, many studies have been reported in recent years concerning brain factors and Alzheimer's disease. Some have identified positive factors for Alzheimer's, such as the loss of myelin (the coating around nerve cells) during middle age. Other studies demonstrate that there are, for example, certain proteins that prevent brain plaques that have been linked to Alzheimer's. At the same time, more and more research emphasizes behaviors – especially sustained active learning activities – that may prevent Alzheimer's. Thus, if Alzheimer's can be facilitated by and prevented by the structures of communities of practice, then the brain is not a simple autonomous site of health and disease. It is a relatively short leap to the conclusion (speculative or hypothetical) that the "community body" is the locus of health and disease and furthermore of thought and consciousness.

Conclusion: Where We Have to Stand in Order to Begin

In writing this book, we situate ourselves in the midst of an on-going revolution that is changing the structure of inquiry, our inquiring practices, and the way we think. Contemporary sociology, anthropology, and science studies reflect and have helped to manufacture this revolution. There are roughly two points of origin for this revolution, which Restivo (1994) has elsewhere called a Copernican social science revolution: the first point of origin crystallizes in the 1840s and gives us classical social theory; the second point of origin which has roots in the first crystallizes in the middle years of the twentieth century. Our generation came of intellectual age in an era of postmodernist discourse. Postmodernism tends to serve – for all of its variety – as the generic term for the intellectual agendas of the last half of the twentieth century. It stands – or can be made to stand – for a recognition of the profound complexity of the world. It has made many of us cautious and even overly cautious about Grand Theories, Grand Narratives, and absolutes and universals of all kinds. The essence of postmodernism may be that it "unstably describes instability" (Boisvert, cited in Dyens 2001, 110).

Similarly, there have been two multicultural revolutions during the last two centuries. The first was wrought by the engagements between peoples and cultures around the world that fashioned east and west. This revolution was fueled by western movements in the provinces of the "exotic" and "savage" Other. The second multicultural revolution during the second half of the twentieth century was fueled by a more self-possessed Other moving into the landscape of a modern world dominated by western economies and technologies. Of course, it moved already transformed by the west into a west already transformed by the east (see the discussion of science and orientalism in Restivo and Loughlin 2000). The pluralities that emerged out of the first revolution were multiplied many-fold times and strengthened by new levels of self-, ethnic-, and cultural consciousness. In the face of the growing awareness of the seemingly endless variety of ways of living and thinking, intellectuals were practically forced to find in this variety a common denominator that reduced them all to or reinvented them as "stories." Inevitably, science was caught in this net and became for many just another story, or a story period since stories were the only strategies available for telling ourselves about our selves and our world(s).

Postmodernisms *in extremis* led to an out of control skepticism about universals, truth, objectivity, and rationality. Relativisms were resurrected across the intellectual landscape. In spite of claims to the contrary by careless and uninformed critics, none of the leading pioneers in science studies defended an "anything goes, everything is equal" relativism. From David Bloor (1999), who defined himself as a champion of western culture and science to Bruno Latour, who apologized in 1987 for the earliest relativistic excesses of the science studies community and recently described himself in realist terms (Latour 2004), none undermined the classical scientific project. In the wake of the excesses one would expect in liminal times, it has become necessary for those of us unintimidated by these excesses to learn again how to tell the truth (Smith 1999).

The variety of postmodernist and poststructuralist excesses goes hand in hand with constant efforts to resurrect the agent that structural approaches tend to keep eliminating. There are many resources brought to bear on the project of resurrecting and sustaining the agent, among them chaos theory, self-organization theory, information theory, genetic theories, and the recurring rehabilitation of transcendence and immanence as acceptable intellectual perspectives. When agency is championed, the sort of sociology we advocate is challenged. Under such conditions it is no wonder that rational choice theory can continue to show up in contemporary sociological theory. We are engaged in a continuing effort to save truth, objectivity, and yes, science from these various excesses, including the excess of trying to dethrone Grand Narratives and Grand Theories using thinly disguised Grand Narratives. When we say we want to save truth, objectivity, and science, we do not mean that we want to save them in their original forms. To save science means to save thinking and inquiry, not the modern social institution of science. Even the liberating celebration of the local can become a "new kind of globalizing imperative."

At the heart of the matters we have addressed, perhaps, is the resistance to the discovery that we human beings are social through and through. There is widespread *cultural* resistance to this idea and to virtually all the great ideas of sociology and anthropology. Many people, many thinkers, and incredibly, many sociologists continue to conceive of individuals as individuals in the strong sense. Furthermore, there is a *methodological* resistance. Just as Kelvin, for example, resisted Maxwell's theory of light because he could not model it mechanically, so many people resist the idea of social construction because they cannot model it after their own fashion and especially because their very selves – *their bodies* – resist it (Butler 1993).

There may also be a sort of *religious* resistance to social science because – despite Saint-Simon, Comte, and even Durkheim himself – sociologists and anthropologists analyze God rather than worship God. And finally, the various forms of *social* resistance play their part – the *relatively low professional standing* of the discoverers, for example; Durkheim and Marx are virtually invisible in the overdrawn shadows of Newton, Galileo, and Einstein; the *prevailing patterns of specialization* that make physicists and astronomers experts on God and souls and spirit instead of anthropologists and sociologists; and then a host of *organizational and structural* factors from age grading to schools and patterns of authority reveal the truth in

T. H. Huxley's (cited in Bibby 1960, 18) remark that "authorities, disciples, and schools are the curse of science." According to Bernard Barber (1990, 111), "As persons in society, scientists are sometimes the agents, sometimes the objects of resistance to their own discoveries."

"Nothing," Nietzsche wrote in *Dawn* (1974, in aphorism 18), "has been purchased more dearly than the little bit of reason and sense of freedom which now constitutes our pride." That little bit of reason and that sense of freedom are not only dearly bought, they are also easily lost. The struggle for reason is a never ending one, but it is especially in liminal times that we must be most vigilant and prepared for struggles of conscience if not for survival. We have wandered into a landscape of complexities that has changed and challenged first our classical systems of classification and our categories and now the very fabric of our cultural inheritances. The urgencies in this landscape – some challenging our very survival as a species – have pressed us to embrace new classifications and categories. This is how in fact we will in the end answer the liminal challenge and construct new logics and rationalities. This is at the same time a process that easily and necessarily leads to excesses and the rise of irrationalities masquerading as the appropriate strategies for the post-liminal period (Restivo 1983). We can embrace new modes of inquiry, thinking, and knowing too soon and with too much certainty. And indeed we have done just this in our era of post-hyphenisms. It is no wonder that when our logics and rationalities fail us, when our *words* fail us, that the world(s) they re-present inevitably fall(s) away and we find ourselves denying truth, objectivity, and reality and doing it with the passion of an Archimedean Eureka! It is useful to consider starting over and adopting the entry strategies of the skilled anthropologist or the historian.

Let us consider, then, the challenge put forward by Foucault (1984):

> I think that the central issue of philosophy and critical thought since the eighteenth century has always been, still is, will, I hope, remain the question: What is this reason that we use? What are its historical effects? What are its limits and what are its dangers? How can we exist as rational beings, fortunately committed to practicing a rationality that is unfortunately criss-crossed with intrinsic dangers? One should remain as close to this question as possible, keeping in mind that it is both central and extremely difficult to resolve. (249)

Similarly, how can we live with institutions that are so obviously productive yet fallible? We are suspicious of the state, knowing full well that it can be captured by powerful and rich people with influence. But we also have no other institutions currently available on a sufficient scale to counter the excesses of post-industrial capitalism. How can we encourage people to love the world around them and to get to know it without reifying that knowledge and transforming it into stale lists of facts, dogmatic worldviews, abuses of certainty, and claims to political power? How can we encourage people to be systematic, critical, and analytic, without settling into a mechanistic system for grinding away at the world? We leave these as open questions for our students, but encourage them to read some recent statements on the nature, history, and future of science studies (Bloor 1999 and the exchange with Latour in the same issue; Latour 2003; Restivo 2004; Latour 2004).

Our stress on sociology and social construction is at odds with at least some trends in science and technology studies. We understand the rationales behind theories that herald the "end of the social" and we are even sympathetic to some of these efforts just because they reflect complexities that challenge long standing disciplinary approaches and divisions in intellectual life. We leave this question too to our readers and to those theorists who are working hard to refashion our modes and methods of inquiry (for one challenging example of such an effort, see Lash 2002). It has not been our goal to privilege sociology as a discipline but rather to privilege a certain form of life and a certain worldview. Sociology may very well go the way of natural philosophy, but the idea of the social is no more likely to disappear than the idea of the material world. In the immediate future, hybrids will abound as humans seek to settle the cultural disequilibria that rocked the twentieth century and try to construct forms of life and knowing that work on a global scale. Science and technology studies is both a reflection of hybridization and an effort to construct a new rationality.

Note

1   There are a number of general reviews of brain research that illustrate the ways in which contemporary neuroscience is discovering, incorporating or otherwise addressing social and cultural factors impacting the brain's structure and connectivities. See, for example, Restak 2003; Czerner 2001; and Kotulak 1997; see also Howe 1999. Elizabeth Wilson's Neural Geographies (1998) is recommended for the more advanced reader. On the sociology of emotions, see Kemper (1990) and other contributions in the SUNY Press series on the sociology of emotions, and Barbalet (2003).

Further Reading

Breazeal, C. (2002). *Designing Sociable Robots*. Cambridge, MA, Mit Press.

Brooks, R. (2002). *Flesh and Machines: How Robots Will Change Us*. New York, Pnatheon.

Brothers, L. (2001). *Mistaken Identity: The Mind – Brain Problem Reconsidered*, Albany, New York, SUNY.

Ginsburg, F. D. and Rapp, R. (eds.) (1995). *Conceiving the New World Order: The Global Politics of Reproduction*. Berkeley, University of California Press.

Jasanoff, S. (1997). *Science at the Bar: Law, Science, and Technology in America*. Cambridge, MA, Harvard University Press.

Mitchell, R. and Thurtle, P. (eds.) (2004). *Data Made Flesh: Embodying Information*. New York, Routledge.

Glossary

androcentric: Dominated by or emphasizing masculine interests or a masculine point of view.

artifacts: Objects and things created and used by humans. Within STS it signifies that the author is giving attention to "the meaning of the designs and arrangements" of the object within a particular environment, practice and/or culture as well as thinking of the object as a tool or practical object (Winner 1986, 25).

autonomous science: Classically, the idea that science is independent of society, history, and culture. This assumption grounded the old sociology of science, which assumed a-social system of science that led to scientific facts free of social and cultural influences (cf. Mannheim's sociology of knowledge and Mertonian sociology of science).

autonomous technology: The common belief that technology is "out of control and follows its own course, independent of human directions" (Winner 1977, 13). This belief is critically discussed and its cultural sources identified in Winner (1977). It is also more generally the idea that technology unfolds (develops, or evolves) in history according to its own inner logic.

closure: The point at which all of the social groups working (or competing) to define a new concept or artifact come (or are forced) to agree on what the new "thing" is.

communalism: One of the four basic norms of science that govern the rules by how scientists work, as proposed by Robert K. Merton (1973). It is the free sharing of information within the scientific community for the advancement of scientific knowledge.

communities of practice: A group of people who share common interests, terminologies, and backgrounds working together to manage the creation of facts and artifacts and to sustain the tacit aspects of work, craft, creation, discovery, and innovation in their community.

convivial society: An alternative society characterized by individual freedom realized by emphasizing the intrinsic value of personal inter- and independence. This society would foster creativity, community and the ability of individuals to create their own futures through power sharing. "A convivial society would be the result of social arrangements that guarantee for each member the most ample and free access to the tools of the community and limit this freedom only in favor of another member's equal freedom" (Illich 1973, 13). Illich uses this concept to analyze and criticize the limitations of industrial society whose logic is based upon addiction to the goods and services of mass production.

culture: The "glue" of shared ideas, manners, and things that unites a community; the totality of a group's ways of living, knowing, and believing. It includes material and non-material aspects from artifacts and things to symbols, attitudes, beliefs, values, norms, and social organization. The anthropologist David Bidney (1967) once labeled the ingredients of culture as artifacts, mentifacts, and socifacts.

cyborg: Proposed in 1960 by Manfred Clynes and Nathan Kline to name the "exogenously extended organizational complex functioning as an integrated homeostatic system unconsciously" (Clynes and Kline 1995: 31). However, since then it has been adopted and adapted in many communities of practice from science fiction to cultural studies. In STS, the concept of Cyborg was introduced by Donna Haraway in "Manifesto for Cyborgs" published in 1985. Haraway (1991, 149) used cyborg to describe "a cybernetic organism, a hybrid of machine and organism, a creature of social reality as well as a creature of fiction." She suggests that we are all cyborgs and by understanding this we can begin to take pleasure in the confusion of the boundaries and assume responsibility in making the boundaries.

disinterestedness: One of the four basic norms of science that govern the rules by how scientist work, as articulated by Robert K. Merton (1973). Scientists are to work for the "best" interests of science and not for themselves. They are to be objective and neutral in their daily practice and ultimately they will be rewarded for their truth seeking.

ethical creep: The concept that certain limited policies or activities carry the potential for introducing larger-scale value changes into a community or society. For example, some critics of cloning think that it is "the end product of an ethical creep of alternative lifestyles and deteriorating fami-lies and promises confusing new relationships resulting from single-parent and lesbian-couple reproduction" (Meyer 1999). It reflects the difficulty in articulating criticisms of technologies that make new activities possible, and represents the collapse of "can" and "should" in ethical thinking.

ethnocentric: Judging or evaluating a culture in whole or in part based upon criteria from one's own culture. The attitude that one's own culture rightly serves as the measure of all cultures.

gemeinschaft–gesellschaft: Polar ends of a continuum that describes the nature of societies and social relationships (Ferdinand Tönnies 1887/1963.) On the gemeinschaft end are social relationships generally based upon intimacy, cooperation and kinship. Gesellschaft refers to social relationships founded upon self-interest and competition. It is kin to a set of dichotomies that set modern urban and industrial life in opposition to traditional ways of living (e.g., urban–rural, city–community, modern–traditional).

glocal: A term used to express the coming together and interpenetration of global and local phenomena. It is a way of stressing the difficulty, even the impossibility, of separating global and local phenomena.

hegemony: Used by Antonio Gramsci (1971) to describe a form of social dominance that is stable, has the support of the dominated, and is accepted as legitimate based on the belief that the ruling elite is just, rational, and working for the common good. It refers to the dominance, the social mechanisms, institutional framework, and social structures that normalize the dominance.

information technology: Any kind of system that can create, convert, transmit, duplicate, or use data. Data is sometimes used as a synonym for information. Information and information technologies embody their social and cultural contexts-of-use. The contemporary ubiquity of com-puters, satellites, and personal portable information devices has helped to fuel the notion that we are in the midst of an information revolution.

institution: A conventional social arrangement that is self-policing because all individuals are committed through a common interest to maintaining it; a "legitimizing social grouping" like the family, religion, or a game. Institutions encode information and provide routine ways of thinking, solving problems, and decision making (Douglas 1986, 46–47). They are stable sets of social relationships that organize the fundamental problem-solving mechanisms of a society or community (mechanisms for dealing with such things as feeding people, clothing and housing them, and protecting social and cultural boundaries).

materialism: Has two meanings, the first being the cultural value that is placed upon the possession of material goods as the bases for identity, status and comfort. The second is a theoretical approach that identifies economic production and reproduction as the foundation of all social life, social change and social development. "Economic" here is understood to mean the identification, exploitation, mobilization, production, distribution, and utilization of resources.

mode of production: The term associated with Karl Marx's way of periodizing history in broadly economic terms: primitive communism, Asiatic, ancient, feudal, and capitalist. Mode of production is characterized according to relation between productive forces and relations of production. For example, the modern bourgeois (capitalistic) mode of production is associated with the conflictful relationship of workers (proletariat) and capitalists (bourgeoisie). Workers do not own the means of production, so must sell their labor to capitalists; the capitalist is destined to pursue the accumulation of capital. Each era has its modal forms of economic organization and social order.

nation-state: A society in a particular territory with a common identity and history (a nation) that is governed by an authority – (a state) – whose boundaries are coincident with those of the nation (cf. Giddens 1985). It represents a way of organizing identity and productive activities, and may be decentered with the advent of contemporary globalization.

nihilism: A philosophy of denial of all established authority and institutions (this philosophy stems from a revolutionary doctrine that supports the destruction of the social system for its own sake). In the extreme, it is an opposition to everything including life itself.

nominalism: The belief that "universals" do not exist, and are merely terms created to describe objects.

objectivity: Objectivity has three connotations. One is based on a sense of neutrality, requiring judgments based on observation and not personal prejudices or opinion. A second sense reflects an understanding of methods being systematic and not idiosyncratic: a scientific technique must be reproducible by any reasonably trained practitioner. The third sense reflects an understanding of something being objective has being real and not subject to dispute (see Daston 1992). For science studies scholars including Traweek and Restivo, terms like "cultures of objectivity" and "objectivity communities" underscore the fact that objectivity is an achievement of social groups, attributed to statements or things through various social processes.

organized skepticism: Is one of the four basic norms of science that govern the rules by how scientists work, as articulated by Robert Merton (1973). It states that scientists should always question and resist the declaration that anything is once and for all "true." For scientists, critical thinking and questioning should be the ever-present foundation of scientific inquiry, and that this skepticism is systematic.

post-hyphen: A generic term coined by Sal Restivo for all things post-, e.g., postmodern, post-Enlightenment, post-industrial, post-human, post-historical, poststructural, especially as these terms have come into general use in the mid- to late twentieth century. It reflects the point at which a concept or term is recognized for having multiple meanings over time and becomes an object of study.

postmodernist: Someone who subscribes to a family of critical ways of thinking about "modernity" and its institutions, including science. A postmodernist might either give up the idea of "truth" or see postmodernism as a pathway to a more sophisticated culturally grounded concept of truth than characterized modernism.

realism: Realism is a viewpoint opposed to nominalism or idealism, and states that things known as "universals" exist independently of society and the material world.

reification: Transforming an idea, concept, or product of the imagination into a material thing (as a matter of belief and usage).

relativism: The position that all criteria of judgment are relative to the individuals and situations involved and that ethical truths depend on the individuals and groups holding them. This includes the belief that all points of view are equally valid. Relativists believe every framework (e.g., culture, individual) has its own relative viewpoints, but that no framework is, can be, or should be dominant or superior. In addition the theory states that knowledge is relative to the limited nature of the mind and the conditions of knowing. In recent sociology of science, relativism has been mistakenly opposed to realism. As a result it has been used to affirm or critique social constructionism. It is more accurately understood as the opposite of absolutism. Barry Barnes and David Bloor (1982, 47), defined relativism as "disinterested inquiry," a classical definition of science.

science wars: The "science wars" emerged in the 1990s as physical and natural scientists began to speak out against what they characterized as the threat to reason and objectivity posed by social and cultural theorists of science. The attacks on the sociological study of science, exemplified by the book *The Higher Superstition* (Gross and Levitt 1994) and the so-called Sokal affair, were based on misreadings and distortions of key ideas and perspectives in science studies.

self-organization theory: A cross-disciplinary approach grounded primarily in physics, chemistry, and molecular biology to analyzing and understanding the nature, emergence, and evolution or development of complex dynamic systems; the theory of how order emerges out of disorder, of the "spontaneous" emergence of form. It is sometimes a generic term for a set of ideas ranging from autopoiesis and self-referential systems to dissipative structures and chaos theory.

semiotics: The study of signs and symbolic systems and their application in everyday life.

social construction: The theory that addresses the inherent social nature of every facet of reality. It outlines the processes of social interactions that give rise to a society's symbols, artifacts, ideas, and behaviors. Social construction focuses on the social group, not the individual. It illustrates that apparently naturalistic or objective phenomena are the outcomes of social processes. It does not deny a "reality out there," but stresses that we can only "know" about that reality through descriptions that are socially and culturally created.

social construction of technology: The idea that technology is socially constructed and that there is not an intrinsic logic of technology. It is a specific school which was originally inspired by Harry Collins' (1985) empirical programme of relativism (EPOR), and research on the social construction of science. Technology and technological systems are analyzed in terms of such concepts as interpretative flexibility, relevant social groups (core groups), and closure.

social webs of meaning: Refers to the fact that all symbols, ideas, concepts, and words in a given social or cultural milieu are variously interconnected and shared by participants in the milieu. More complexly, their meanings flow into each other to different degrees (cf. Mary Hesse 1974; Barnes 1983, 23ff).

stratification: Refers in general to social, religious, political, or economic hierarchies; each level in the hierarchy is characterized by it position and power relative to the other levels. For example, in feudal societies, the hierarchy places slaves and peasants on the bottom, followed by knights, vassals, princes/princesses, and kings/queens). Stratification can be based on different dimensions, corresponding to different realms of social activity. Technoscience is believed to be a meritocracy – a system stratified by ability or merit. A plutocracy is a social system based on wealth.

technique: Used by Jacques Ellul (1964, xxv) to describe the "totality of methods rationally arrived at and having absolute efficiency in every field of human activity." According to Ellul it is the new milieu of human activity and has replaced nature as humans' day-to-day environment. Technique is artificial, autonomous, self-determining, objective, and interrelated. Technique attends all social activity and makes interactions instrumental or valuable only in their measurable benefit to an actor.

technological determinism: The idea, strongly criticized in science and technology studies, that technologies follow trajectories of development whose logic is innate and not tied to social and cultural factors; and that technology's impact or influence on society is unidirectional.

technological fix: The belief that social problems (like technological problems) can be solved by the application of technological reasoning. This belief assumes a separation between "technology" and "society" and thus violates basic assumptions in STS including the concepts of the technosocial and technoscience. It tries to deal with the complexity of social problems by identifying some small piece of the problem amenable to technological solution.

technological flexibility/affordance: Technologies can and are used differently than their inventors and designers intended or anticipated. Affordance is a term reflecting designed features that facilitate the use of a technology for certain activities, while flexibility represents the idea that affordances can apply to multiple activities and objectives. It shows technologies are originally designed for a certain purpose and that design is a factor limiting their arbitrary use, while technologies might also be used for and bring about unintended consequences. In sum, a technology has flexibility to make different outcomes from initial intentions, however it should be understood that things have physical and contextual limitations in use and adaptation.

technological intensification: The process where social phenomena intensify as people select technologies. It is used to clarify the logical and real sequences of technological development and social change. It is a response to the ideas that technology drives history and that the social somehow must respond to technological change by emphasizing that the social factors that go into the invention and selection of technology also contribute to its effects.

technoscience: This concept variously refers to the idea that science and technology are intricately interrelated and that the separation of science and technology is analytical at best.

technosocial: Conceptualizing technologies as embodying social dimensions. See technoscience.

Turing test: Devised by Alan Turing (1950). It is an "imitation game" in which a human questioner would interrogate either a human being or a computer by textual messages. The questioner would not know who was replying and would be required to identify the responses as originating from either a human or a computer. Turing argued that if the interrogator could not distinguish the human from the computer through questioning, then it would be reasonable to call the computer intelligent.

universalism: Is one of the four basic norms of science that govern the rules by how scientists work, as articulated by Robert Merton (1973). This norm states that scientists should evaluate scientific findings objectively based upon the experimental and theoretical merit and not on subjective criteria of the scientists, or the attributes (such as race, class, gender, age) of the person proposing the knowledge claim.

worldview: A general and comprehensive way of seeing, interpreting and relating to the world from a cultural perspective that provides the tools to categorize and classify experience. It is an all-encompassing framework or a perspective composed of shared meanings, values and explanations.

Acknowledgment

This glossary was prepared with the assistance of graduate students Byoungyoon Kim, Toluwalogo B. Odumosu, Olivette T. Sturges, David W. vonEiff, and undergraduate senior George D. Orfanos from the Science and Technology Studies Department at Rensselaer Polytechnic Institute, Troy New York.

References

A. R. T. Law "Frozen Embyro Custody." Retrieved March 2003 from http://infertility.about.com/library/weekly/aa041500a.html.

Abraham, C. (2001). *Possessing Genius: The Bizarre Odyssey of Einstein's Brain*. New York, St. Martin's Press.

Adas, M. (1989). *Machines as the Measure of Men: Science, Technology, and Ideologies of Western Dominance*. Ithaca, NY, Cornell University Press.

Addelson, K. (1983). "The Man of Professional Wisdom" In *Discovering Reality*. S. Harding and M. Hintikka (eds.). Dordrecht, Holland, D. Reidell: 165–186.

Akrich, M. (1992). "The De-Scription of Technical Objects." In *Shaping Technology/Building Society: Studies in Sociotechnical Change*. W. E. Bijker and J. Law (eds.). Cambridge, MIT Press: 205–224.

Allen, S. G. and J. Hubbs (1987). "Outrunning Atlanta: Feminine Destiny in Alchemical Transmutation." In *Sex and Scientific Inquiry*. S. Harding and J. F. O'Barr (eds.). Chicago, University of Chicago Press: 79–98.

Althusser, L. (1998). "Ideology and the Ideological State Apparatuses." In *Cultural Theory and Popular Culture: A Reader*. J. Storey (ed.). Athens, University of Georgia Press: 153–164.

Andersen, J. (2001). "The Status of Pluto: A Clarification." International Astronomical Union. Retrieved 3/01/04 from http://www.iau.org/IAU/FAQ/PlutoPR.html.

Aronowitz, S. (1988). *Science as Power: Discourse and Ideology in Modern Science*. Minneapolis, University of Minnesota Press.

Astington, J. (1996). "What is Theoretical About the Child's Theory of Mind?: A Vygotskian View of its Development." In *Theories of Theories of Mind*. P. Carruthers and P. K. Smith (eds.). Cambridge, Cambridge University Press.

Avila, D. *Massachusetts Court Rules in Frozen Embryo Case.* Retrieved 6/10/04 from http://www.nrlc.org/news/2000/NRL04/mass.html.

Bakhtin, M. M. (1981). *The Dialogic Imagination: Four Essays.* Austin, University of Texas Press.

Bakhtin, M. M. (1986). *Speech Genres and Other Late Essays.* Austin, University of Texas Press.

Barad, K. (2003). "Posthuman Performativity: Toward an Understanding of How Matter Comes to Matter." *Signs: A Journal of Women and Culture,* 28(31): 801–831.

Barbalet, J. (ed.) (2003). *Emotions and Sociology.* London, Blackwell.

Barber, B. (1990). *Social Studies of Science.* New Brunswick, Transaction Publishers.

Barber, B. (2002). "Beyond Jihad vs McWorld: On Terrorism and the New Democratic Realism." *The Nation,* 274(2): 11–18.

Barnes, B. (1983). "On the Conventional Character of Knowledge and Cognition." In *Science Observed.* K. Knorr-Cetina and M. Mulkay (eds.). Beverly Hills, CA, Sage: 19–51.

Barnes, B. and D. Bloor (1982). "Relativism, Rationalism and the Sociology of Knowledge." In *Rationality and Relativism.* M. Hollis and S. Lukes (eds.). Cambridge, MIT Press: 21–47.

Barnes, B. and S. Shapin (eds.) (1979). *Natural Order: Historical Studies of Scientific Culture.* Beverly Hills, Sage Publications.

Baron, J. B. and J. Epstein (1998). "Language and the Law: Literature, Narrative and Legal Theory." In *Politics of Law: A Progressive Critique.* D. Kairys. New York, Basic Books.

Basalla, G. (1974). "The Spread of Western Science." In *Comparative Studies in Science and Society.* S. Restivo and C. K. Vanderpool (eds.). Columbus, Charles E. Merrill: 359–381.

Basalla, G. (1988). *The Evolution of Technology.* Cambridge, Cambridge University Press.

Beck, U. (2000). *What Is Globalization?* London, Polity Press.

Becker, H. (1982). *Art Worlds.* Berkeley, University of California Press.

Becker, H., and M. McCall (1990). *Symbolic Interaction and Cultural Studies.* Chicago, IL, University of Chicago Press.

Benedict, R. (1934). *Patterns of Culture.* Boston, Houghton Mifflin Company.

Benjamin, W. (1968). "The Work of Art in the Age of Mechanical Reproduction." In *Illuminations.* H. Arendt (ed.). New York, Harcourt, Brace & World: 219–254.

Berger, P. (1963). *Invitation to Sociology.* New York, Anchor Books.

Biagioli, M. (1996). "From Relativism to Contingentism." In *The Disunity of Science.* P. Galison and D. J. Stump (eds.). Stanford, Stanford University Press: 189–207.

Bibby, C. (1960). *T.H. Huxley: Scientist, Humanist and Educator.* New York, Horizon Press.

Bidney, D. (1967). *Theoretical Anthropology,* 2nd augmented edn. New York, Schocken Books.

Bijker, W. E. (1995). *Of Bicycles, Bakelites, and Bulbs: Toward a Theory of Sociotechnical Change.* Cambridge, MIT Press.

Bijker, W. E. (2001). "Understanding Technological Culture Through a Constructivist View of Science, Technology, and Society." In *Visions of STS: Counterpoints in Science, Technology, and Society Studies.* S. H. Cutliffe and C. Mitcham (eds.). Albany, New York, SUNY Press, 19–34.

Bijker, W. E., T. P. Hughes, and T. J. Pinch (eds.) (1987). *The Social Construction of Technological Systems: New Directions in the Sociology and History of Technology.* Cambridge, MA, MIT Press.

Birke, L. (2001). "The Pursuit of Difference: Scientific Studies of Women and Men." In *The Gender and Science Reader.* Muriel Lederman and Ingrid Bartsch (eds.). New York, Routledge, 309–322.

Blakeslee, S. (1991). "Brain Yields New Clues on its Organization for Language." *New York Times,* Sept. 10: C1.

Bleier, R. (2001). "Sociobiology, Biological Determinism, and Human Behavior." In *Women, Science, and Technology: A Reader in Feminist Science Studies.* Mary Wyer, Mary Barbercheck, Donna Giesman, Hatice Örün Öztürk, and Marta Wayne (eds.). New York, Routledge: 175–194.

Blizzard, D. (2005). *Looking Within: A Socio-cultural Examination of Fetoscopy.* Cambridge, MA: MIT Press.

Bloch, H. (1972). "The Problem Defined." In *Civilization & Science: In Conflict or Collaboration?* Ciba Foundation. New York, Elsevier: 1–7.

Bloor, D. (1991). *Knowledge and Social Imagery.* Chicago, University of Chicago Press.

Bloor, D. (1999). "Anti-Latour." *Studies in History and Philosophy of Science,* 30(1): 81–112.

Booth, M. (2001). *New Jersey High Court Sets Framework for Custody and Use of Frozen Embryos, New Jersey Law Journal.* Retrieved 2003. http://www.law.com/jsp/statearchive.jsp?type=Article&oldid=ZZZ0NCK4HQC.

Bordo, S. (1987). "The Cartesian Masculinization of Thought." In *Sex and Scientific Inquiry.* S. Harding and J. O'Barr (eds.). Chicago, University of Chicago Press: 247–264.

Bowker, G. C. and S. L. Star (1999). *Sorting Things Out: Classification and Its Consequences.* Cambridge, MIT Press.

Boyer, C. (1968). *A History of Mathematics.* New York, John Wiley & Sons.

Breazeal, C. (1999). "A Context Dependent Attention System for a-Social Robot." In *Proceedings of the Sixteenth International Joint Conference on Artificial Intelligence,* Stockholm, Sweden.

Breazeal, C. (1999). "Robot in Society: Friend or Appliance?" In *Agents99: Workshop on Emotion-Based Agent Architectures,* Seattle, WA: 18–26.

Briggs, H. (2004). "New Planet Forces Rethink." BBC News, World Edition. Retrieved 4/2/04 from http://news.bbc.co.uk/2/hi/science/nature/3516952.stm.

Brooks, R. (1999). *Cambrian Intelligence.* Cambridge, MIT Press.

Brothers, L. (1997). *Friday's Footprint: How Society Shapes the Human Mind.* New York, Oxford University Press.

Brothers, L. (2001). *Mistaken Identity: The Mind–Brain Problem Reconsidered*. Albany, SUNY Press.

Butler, J. (1993). *Bodies that Matter: On the Discursive Limits of Sex*. New York: Routledge.

Byrne, R. W. (1991). "Brute Intellect." *The Sciences*: 42–47.

Campbell, D. T. (1964). "Distinguishing Differences of Perception from Failures of Communication in Cross-Cultural Studies." In *Cross-Cultural Understanding: Epistemology in Anthropology*. F. S. C. Northrop and H. H. Livingston (eds.). New York, Harper and Row: 308–338.

Castells, M. (1998). *The Information Age: Economy, Society and Culture*. Oxford, Blackwell.

Clarke, A. (1998). *Disciplining Reproduction: Modernity, American Life Science and the "Problem of Sex."* Berkeley, University of California Press.

Clarke, A. and J. H. Fujimura (eds.) (1992). *The Right Tools for the Job: At Work in 20th Century Life Sciences*. Princeton, Princeton University Press.

Clynes, M. E. and N. S. Kline (1995). "Cyborgs and Space." In *The Cyborg Handbook*. C. H. Gray (ed.). New York and London, Routledge: 29–35.

Cockburn, C. and S. Ormrod (1993). *Gender and Technology in the Making*. London, Sage.

Collins, H. M. (1985). *Changing Order: Replication and Induction in Scientific Practice*. London, Sage.

Collins, H. M. (1990). *Artificial Experts: Social Knowledge and Intelligent Machines*. Cambridge, MA, MIT Press.

Collins, H. M. (1998). *The Shape of Actions: What Humans and Machines Can Do*. Cambridge, MA, MIT Press.

Collins, H. and T. Pinch (1998a). *The Golem at Large: What You Should Know About Technology*. Cambridge, Cambridge University Press.

Collins, H. and T. Pinch (1998b). *The Golem: What You Should Know About Science* (2nd edn.). Cambridge, Cambridge University Press.

Collins, R. (1998). *The Sociology of Philosophies: A Global Theory of Intellectual Change*. Cambridge, MA, London, Belknap Press of Harvard University Press.

Collins, R. and M. Makowsky (1998). *The Discovery of Society* (6th edn.). New York, McGraw Hill.

Collins, R. and S. Restivo (1983). "Robber Barons and Politicians in Mathematics." *Canadian Journal of Sociology*, 8: 199–227.

Cowan, R. S. (1989). *More Work for Mother: The Ironies of Household Technology from the Open Hearth to the Microware*. London, Free Association Press.

Croissant, J. (2000). "Engendering Technology Culture, Gender and Work." *Knowledge and Society*, 12, 189–207.

Croissant, J. and S. Restivo (1995). "Science, Social Problems, and Progressive Thought." In *Ecologies of Knowledge: Work and Politics in Science and Technology*. S. L. Star (ed.). Albany, SUNY Press: 39–88.

Cutcliffe, S. and C. Mitcham (eds.) (2001). *Visions of STS: Counterpoints in Science, Technology, and Society Studies*. Albany, SUNY Press.

Czerner, T. B. (2001). *What Makes You Tick? The Brain in Plain English.* New York, John Wiley & Sons.

Daar, J. F. (1999). "Assisted Reproductive Technologies and the Pregnancy Process: Developing an Equality Model to Protect Reproductive Liberties." *American Journal of Law and Medicine,* 25(4): 455–477.

Damasio, A. R. (1994). *Descartes' Error: Emotion, Reason, and the Human Brain.* New York, G. P. Putman.

Daston, L. (1992). "Objectivity and the Escape from Perspective." *Social Studies of Science,* 22: 597–618.

Dickson, D. (1974). *Alternative Technology and the Politics of Technical Change.* London, Fontana.

Douglas, M. (1986). *How Institutions Think.* Syracuse, Syracuse University Press.

Douglas, M. (1992). *Risk and Blame: Essays in Cultural Theory.* New York: Routledge.

Drori, G., J. H. Meyer, F. O. Ramirez, and E. Schofer (eds.) (2003). *Science in the Modern World Polity: Institutionalization and Globalization.* Stanford, CA: Stanford University Press.

Durkheim, E. (1961). *The Elementary Forms of the Religious Life.* New York, Collier Books.

Durkheim, E. (1995). *The Elementary Forms of Religious Life.* New York, Free Press.

Dyens, O. (2001). *Metal and Flesh.* Cambridge, MIT Press.

Ellul, J. (1964). *The Technological Society.* New York, Vantage Books.

English-Lueck, J. A. (2002). *Cultures@Silcon Valley.* Stanford, Stanford University Press.

Epstein, S. (1996). *Impure Science: AIDS, Activism, and the Politics of Knowledge.* Berkeley, University of California Press.

Feyerabend, P. (1978). *Science in a Free Society.* London, NLB.

Finn, J. (1964). The Franks Had the Right Idea. *NEAJournal,* 53(4): 24–27.

Fischer, C. (1992). *America Calling: A Social History of the Telephone to 1940.* Berkeley: University of California Press.

Fleck, L. (1935/1979). *Genesis and Development of a Scientific Fact.* Chicago, University of Chicago Press.

Fodor, J. A. (1983). *The Modularity of Mind: An Essay on Faculty Psychology.* Cambridge, MIT Press.

Foucault, M. (1979). *Discipline and Punish.* New York, Vintage Books.

Foucault, M. (1984). "Space, Knowledge, and Power", *The Foucault Reader,* ed. P. Rabinow. New York, Pantheon Books: 239–256.

Fox, M. F. (1999). "Gender, Hierarchy, and Science." In *Handbook of the Sociology of Gender.* J. Saltzman Chafetz. New York, Kluwer/Plenum: 441–457.

Freeman, D. H. (1998). "When is a Planet Not a Planet." *The Atlantic Monthly,* 291(2): 22–33.

Garrett, P. "Endgame: Reproductive Technology & The Death of Natural Procreation." Retrieved 3/03 from http://www.lifeuk.org/speech5.html.

Geertz, C. (1973). *The Interpretation of Cultures.* New York, Basic Books.

Gibbon, E. (1776–1788/1993). *The Decline and Fall of the Roman Empire.* New York: Everyman's Library. Boxed edition (October 26, 1993).

Giddens, A. (1985). *The Nation-State and Violence*. Cambridge, Polity Press.

Ginsburg, F. D. and R. Rapp (eds.) (1995). *Conceiving the New World Order*. Berkeley, University of California Press.

Ginzberg, R. (1989). "Uncovering Gynocentric Science." In *Feminism and Science*. N. Tuana. Bloomington, Indiana University Press: 69–84.

Goodman, E. (1997). "The High Tech 2-Year Old Started with 5 Parents: Now She Has None." *Arizona Daily Star*. 9/16/97, A-10.

Gordon, S. (1985). "Micro-Sociological Theories of Emotion." In *Micro-Sociological Theory: Perspectives in Sociological Theory*. H. J. Helle and S. N. Eisenstadt (eds.). Beverly Hills, Sage: 133–147.

Gorenstein, S. (ed.) (2000). "Enginering Technology: Culture, Gender, and Work." In *Knowledge and Society* vol. 12, 189–207.

Gramsci, A. (1971). *Selections from Prison Notebooks*. London, New Left Books.

Gross, P. and N. Leavitt (1994). *Higher Superstition: The Academic Left and Its Quarrels with Science*. Baltimore, Johns Hopkins University Press.

Hacking, Ian (1983). *Representing and Intervening: Introductory Topics in the Philosophy of Natural Science*. Cambridge, England, Cambridge University Press.

Haraway, D. (1989). *Primate Visions: Gender, Race, and Nature in the World of Modern Science*. New York, Routledge.

Haraway, D. (1991). *Simians, Cyborgs, and Women*. New York, Routledge.

Haraway, D. (1997). *Modest_Witness@Second_Millennium.FemaleMan_Meets_ OncoMouse Feminism and Technoscience*. New York, Routledge.

Harding, S. (1986). *The Science Question in Feminism*. Ithaca, NY, Cornell University Press.

Harding, S. (1991). *Whose Science? Whose Knowledge? Thinking from Women's Lives*. Ithaca, NY, Cornell University Press.

Harding, S. (ed.) (1993). *The "Racial" Economy of Science: Toward a Democratic Future*. Bloomington, Indiana University Press.

Hardt, M. and A. Negri (2000). *Empire*. Cambridge, Harvard University Press.

Heelan, P. (1983). *Space-perception and the Philosophy of Science*. Berkeley, University of California Press.

Heldke, L. (1989). "John Dewey and Evelyn Fox Keller: A Shared Epistemological Tradition." In *Feminism and Science*. N. Tuana (ed.). Bloomington, Indiana University Press: 104–115.

Hess, D. (1997). *Science Studies: An Advanced Introduction*. New York, NYU Press.

Hess, D. (ed.) (1999). *Evaluating Alternative Cancer Therapies: A Guide to the Science and Politics of an Emerging Medical Field*. New Brunswick, Rutgers University Press.

Hess, D. (1995). *Science and Technology in a Multicultural World: The Cultural Politics of Facts and Artifacts*. New York, Columbia University Press.

Hesse, M. (1974). *The Structure of Scientific Inference*. London, Macmillan.

Hilts, P. H. (1991). "Photos Show Mind Recalling a Word." *New York Times*, Nov. 11: A1,8.

Hoffman, R. (1988). "Under the Surface of the Chemical Article." *Angewandte Chemie (International English Edition)*, 27: 1593–1602.

Hooker, C. (1975). *A Realistic Theory of Science*. Albany, SUNY Press.

hooks, b. (1982). *Ain't I a Woman? Black Women and Feminism*. London, Pluto Press.

Horkheimer, M. and T. W. Adorno (1993). *Dialectic of Enlightenment*. New York, Continuum.

Horton, R. (1967). "African Traditional Thought and Western Science." *Africa*, 37(1–2): 50–187.

Howe, M. J. A. (1999). *Genius Explained*. Cambridge, Cambridge University Press.

Ihde, D. and E. Selinger (eds.) (2003). *Chasing Technoscience*. Bloomington, Indiana University Press.

Illich, I. (1973). *Tools for Conviviality*. New York, Harper and Row.

Irigaray, L. (1989). "Is the Subject of Science Sexed?" In *Feminism and Science*. N. Tuana. Bloomington, Indiana University Press: 58–68.

Jacob, E. (1992). "Culture, Context and Cognition." In *The Handbook of Qualitative Research in Education*. M. D. LeCompte, W. L. Millroy and J. Preissle (eds.). San Diego, Academic Press, Inc.: 293–336.

John (1962). *Modern King James Version of the Holy Bible*. New York, McGraw-Hill.

Kaplan, G. and L. J. Rogers (2001). "Race and Gender Fallacies: The Paucity of Biological Determinist Explanations of Difference." In *The Gender and Science Reader*. M. Ledermand and I. Bartsch (eds.). New York, Routledge: 323–342.

Karp, J. and S. Restivo (1974). "Ecological Factors in the Emergence of Modern Science." In *Comparative Studies in Science and Society*. S. Restivo and C. K. Vanderpool (eds.). Columbus, Charles E. Merrill: 123–143.

Keller, E. F. (1985). *Reflections on Gender and Science*. New Haven: Yale University Press.

Keller, E. F. (1989). "The Gender/Science System: or, Is Sex to Gender as Nature Is to Science?" In *Feminism and Science*. N. Tuana (ed.). Bloomington, Indiana University Press: 33–44.

Keller, E. F. (1995). "The Origin, History, and Politics of the Subject Called 'Gender and Science': A First Person Account." In *Handbook of Science and Technology Studies*. S. Jasanoff, G. E. Markle, J. C. Petersen, and T. Pinch (eds.). Thousand Oaks, Sage, 80–94.

Kemper, T. (ed.) (1990). *Research Agendas in the Sociology of Emotions*, Albany, SUNY Press.

Kennedy, J. and R. C. Eberhart (2001). *Swarm Intelligence*. San Francisco, Morgan Kaufmann Publishers.

Kessler, S. (1998). *Lessons from the Intersexed*. New Brunswick, NJ, Rutgers University Press.

Knorr-Cetina, K. D. (1981). *The Manufacture of Knowledge: An Essay on the Constructivist and Contextual Nature of Science*. Oxford, Pergamon Press.

Kotulak, R. (1997). *Inside the Brain: Revolutionary Discoveries of How the Mind Works*. Kansas City, MO, Andrew McMeel Publishers.

Kreisberg, S. (1992). *Transforming Power: Domination, Empowerment and Education*. Albany, SUNY Press.

Kuhn, T. (1962). *The Structure of Scientific Revolutions*. Chicago, University of Chicago Press.

Lakoff, G. and M. Johnson (1980). *Metaphors We Live By*. Chicago, University of Chicago Press.

Lansing, J. S. (1991). *Priests and Prgorammers: Technologies of Power in the Engineered Landscape of Bali*. Princeton, Princeton University Press.

Lash, S. (2002). *Critique of Information*. Thousand Oaks, Sage.

Latour, B. (1987). *Science in Action: How to Follow Scientists and Engineers Through Society*. Cambridge, Harvard University Press.

Latour, B. (June 2003). "The World Wide Lab. Research Space: Experimentation with Representation Is Tyranny. *Wired*, *11*(6). Retrieved 8/9/04 from http://www.wired.com/wired/archive/11.06/research_spc.html.

Latour, B. (2004). "Why Has Critique Run Out of Steam? From Matters of Fact to Matters of Concern." *Critical Inquiry*, 30: 2.

Latour, B. and S. Woolgar (1979, 1986). *Laboratory Life*. Beverly Hills, Sage.

Lave, J. and E. Wenger (1991). *Situated Learning: Legitimate Peripheral Participation*. Cambridge and New York, Cambridge University Press.

Lederman, M. and I. Bartsch (eds.) (2001). The Gender and Science Reader. London and New York, Routledge.

Lemert, C. (ed.) (1993). *Social Theory: The Multicultural & Classic Readings*. Boulder, Westview Press.

Lemonick, M. D. (1997). "The New Revolution in Making Babies." *Time Magazine* 150(23).

Lévi-Strauss, C. (1966). *The Savage Mind*. Chicago, University of Chicago Press.

Lock, M. (1996). "Death in Technological Time." *Medical Anthropology Quarterly*, 10(4): 575–600.

Longino, H. (1990). *Science as Social Knowledge: Values and Objectivity in Scientific Inquiry*. Princeton, NJ, Princeton University Press.

Longino, H. (2002). *The Fate of Knowledge*. Princeton, NY: Princeton University Press.

Loughlin, J. (1993). "The Feminist Challenge to Social Studies of Science." In *Controversial Science: From Content to Contention*. T. Brante, S. Fuller, and W. Lynch (eds.). Albany, SUNY Press: 3–20.

Maddox, B. (2002). *Rosalind Franklin: The Dark Lady of DNA*. New York, Harper Collins.

Malinowski, B. (1954/1948). *Magic, Science and Religion and Other Essays*. New York, Doubleday Ancho Books.

Martin, E. (1990). "The End of the Body?" American Ethnological Society Distinguished Lecture, Atlanta, GA.

Martin, E. (1994). *Flexible Bodies: Tracking Immunity in American Culture From the Days of Polio to the Age of AIDS*. Boston, Beacon Press.

Martin, E. (2001). "Premenstrual Syndrome, Work Discipline, and Anger." In *Women, Science, and Technology: A Reader in Feminist Science Studies*. Mary Wyer, Mary Barbercheck, Donna Giesman, Hatice Örün Öztürk, and Marta Wayne (eds.). New York, Routledge: 285–302.

Marx, K. (1959). *The Economic and Philosophic Manuscripts of 1844*. Moscow, Foreign Languages Publishing House.

Marx, K. (1964). *Economic and Philosophic Manuscripts of 1844*. New York, International Publishers.

Marx, K. (1970). "A Contribution to the Critique of Hegel's Philosophy of Right." In *Marx's Critique of Hegel's Philosophy of Right*. J. O'Malley (ed.). Cambridge, Cambridge University Press.

Marx, K. (1974). *Early Writings*. London, Penguin Books.

Marx, K. (1867/1992). *Capital*. New York, Penguin.

Marx, K. and F. Engels (1947). *The German Ideology*. New York, International Publishers.

Marx, L. (1964). *The Machine in the Garden: Technology and the Pastoral Ideal in America*. New York, Oxford University Press.

McGinn, R. (1991). *Science, Technology and Society*. Englewood Cliffs, Prentice Hall.

McNeill, W. (1965). *The Rise of the West*. Chicago, University of Chicago Press.

Mead, G. H. (1934). *Mind, Self, and Society: From the Standpoint of a-Social Behaviorist*. Chicago, University of Chicago Press.

Merton, R. K. (1973). *The Sociology of Science: Theoretical and Empirical Investigations*, ed. N. Storer. Chicago: University of Chicago Press.

Meyer, C. R. (1999). Cloning Of Wonders Wild & New, v. 92. Retrieved from http://www.mnmed.org/publications/MnMed1999/March/Meyer.cfm.

Mills, C. W. (1959). *The Sociological Imagination*. Oxford, Oxford University Press.

Morone, J. G. and E. J. Woodhouse (1986). *Averting Catastrophe: Strategies for Regulating Risky Technologies*. Berkeley, University of California Press.

Mumford, L. (1934). *Technics and Civilization*. San Diego, Harvest.

Mumford, L. (1964). *The Myth of the Machine: The Pentagon of Power*. New York, Harcourt Brace Jovanovich, Inc.

Nader, L. (1996). "The Three-Cornered Constellation: Magic, Science, and Religion Revisited." In *Naked Science: Anthropological Inquiry into Boundaries, Power, and Knowledge*. L. Nadar. New York, Routledge: 250–276.

Needham, J. and R. Temple (1986). "Science and Civilization in China." *The Genius of China: 3000 Years of Science, Discovery, and Invention*. New York, Simon & Schuster.

Nietzsche, F. (1882/1956). *On the Genealogy of Morals*. New York: Anchor Books).

Nietzsche, F. (1882/1974). *The Gay Science*. New York: Vintage Books.

Nietzsche, F. (1888/1956). *The Birth of Tragedy and The Genealogy of Morals*. Garden City, Doubleday.

Nietzsche, F. (1968). *The Will to Power*. New York, Vintage Books.

Nietzsche, F. (1974). *The Complete Works of Friedrich Nietzsche: the First Complete and Authorised English Translation*. New York, Gordon Press.

O'Keefe, D. L. (1982). *Stolen Lightning: The Social Theory of Magic*. New York, Continuum.

Overbye, D. (1993). "Who is Afraid of the Big Bad Bang?" *Time Magazine*, 141: 74.

Pacey, Arnold (1976). *The Maze of Ingenuity: Ideas and Idealism in the Development of Technology*. Cambridge, MA, MIT Press.

Pacey, A. (1983). *The Culture of Technology*. Oxford, B. Blackwell.

Pacey, A. (1998). *Technology in World Civilization*. Cambridge, MA, MIT Press.

Pacey, A. (1999). *Meaning in Technology*. Cambridge, MA, MIT Press.

Paterniti, M. (2000). *Driving Mr. Albert: A Trip Across America with Einstein's Brain*. New York, Dial Press/Random House.

People's Daily (2000). Retrieved from http://fpeng.peopledaily.com.cn/200201/17/eng20020117_88804.shtml.

Perrow, C. (1984). *Normal Accidents: Living with High-Risk Technologies*. New York, Basic Books.

Pert, C. (1997). *Molecules of Emotion: Why You Feel the Way You Feel*. New York, Scribner.

Pickering, A. (1995). *The Mangle of Practice: Time, Agency, and Science*. Chicago, IL, University of Chicago Press.

Pinker, S. (1994). *The Language Instinct: How the Mind Creates Language*. New York: Harper Collins.

Pinker, S. (1997). *How the Mind Works*. New York, W. W. Norton Co.

Pinker, S. and S. Rose, "The Two Steves: Pinker vs. Rose – A Debate (Part I)." Retrieved 2004 from http://www.edge.org/3rd_culture/pinker_rose/pinker_rose_p1.html.

Pratchett, T. (1992). *Small Gods*. New York, Harper Collins.

Rabinow, P. (ed.) (1984). *The Foucault Reader*. New York, Pantheon Books.

Rapp, R. (1990). "Constructing Amniocentesis: Maternal and Medical Discourses." In *Uncertain Terms: Negotiating Gender in American Culture*. F. Ginsburg and A. L. Tsing (eds.). Boston, Beacon Press: 28–42.

Rapp, R. (1999). *Testing Women, Testing the Fetus: The Social Impact of Amniocentesis in America*. New York, Routledge.

Reaves, J. (2000). "When a Couple Divorces, Who Owns the Embryo?" Retrieved 6/2/2000 from http://www.intendedparents.com/News/When_a_couple_divorces_who_owns_the_embryo.html.

Reich, R. B. (1992). *The Work of Nations: Preparing Ourselves for 21st Century Capitalism*. New York, Vintage Press.

Reid, I. S. (1987). "Science, Politics and Race." In *Sex and Scientific Inquiry*. S. Harding and J. F. O'Barr. Chicago, University of Chicago Press: 99–124.

Restak, R. (2003). *The New Brain: How the Modern Age Is Rewiring Your Mind*. New York, St. Martin's Press/Rodale.

Restivo, S. (1983). *The Social Relations of Physics, Mysticism, and Mathematics*. Dordrecht, Kluwer Academic Publishers.

Restivo, S. (1991). *The Sociological Worldview*. Cambridge, MA, Blackwell.

Restivo, S. (1992). *Mathematics in Society and History: Sociological Inquiries*. Dordrecht, Kluwer Academic Publishers.

Restivo, S. (1994). *Science, Society and Values: Toward a Sociology of Objectivity*. Bethlehem, Lehigh University Press.

Restivo, S. (2003b). "Einstein's Brain, Napoleon's Penis, and Galileo's Finger: Toward a Sociology of the Brain." Lecture in the Science Studies Program colloquium series at the University of California at San Diego, November 24.

Restivo, S. (2004). "What is Science?" In *Life: The Science of Biology* (7th edn.). G. H. Orians, D. Sadava, W. K. Purves, and C. Heller (eds.). New York, Sinauer/Freeman.

Restivo, S. and J. Loughlin (July 2000). "The Invention of Science." *Cultural Dynamics*, 12(2): 135–149.

Ritzer, G. (2004). *The McDonaldization of Society*. Revised new century edn. Thousand Oaks, CA, Sage/Pine Forge.

Robertson, R. (1992). *Globalization: Social Theory and Global Culture*. London, Sage.

Robertson, R. (1994). "Globalisation or Blocalisation." *Journal of International Communication*, 1(1): 33–52.

Robertson, R. (2001). "Globalization Theory 2000+ Major Problematics." In *Handbook of Social Theory*. George Ritzer and Barry Smart (eds.). London, Sage: 458–471.

Rochlin, G. I. (1997). *Trapped in the Net: The Unanticipated Consequences of Computerization*. Princeton, Princeton University Press.

Rorty, R. (1987). "Pragmatism and Philosophy." In *After Philosophy*. K. Baynes, J. Bohman, and T. McCarthy (eds.). Cambridge, MA, MIT Press: 26–66.

Rose, H. (1994). *Love, Power and Knowledge: Towards a Feminist Transformation of the Sciences*. Cambridge, Polity Press.

Savan, B. (1988). *Science Under Siege: The Myth of Objectivity in Scientific Research*. Montreal, CBC Enterprises.

Schiebinger, L. (1989). *The Mind Has No Sex? Women in the Origins of Modern Science*. Boston, Beacon Press.

Schiebinger, L. (1993). *Nature's Body: Gender in the Making of Modern Science*. Boston, Beacon Press.

Schiffer, M. B. (with T. C. Butts and K. K. Grimm) (1994). *Taking Charge: The Electric Automobile in America*. Washington DC, Smithsonian Institution Press.

Searle, J. (1984). *Minds, Brains, and Science*. Cambridge, Harvard University Press.

Shapin, S. (1989). "The Invisible Technician." *American Scientist*, 77: 554–563.

Sismondo, S. (2003). *Introduction to Science and Technology Studies*. London, Blackwell.

Sjoberg, G. (1960). *The Pre-Industrial City*. New York, The Free Press.

Smith, D. (1999). *Writing the Social: Critique, Theory, and Investigations*. Toronto, University of Toronto Press.

Smith, M. (1978/1987). *Jesus the Magician*. San Francisco, Harper and Row.

Spengler, O. (1926). *The Decline of the West*. New York, Knopf.

Star, S. L. (1989). *Regions of the Mind: Brain Research and the Quest for Scientific Certainty*. Stanford, CA, Stanford University Press.

Star, S. L. (ed.) (1995a). *Ecologies of Knowledge: Work and Politics in Science and Technology*. Albany, New York, SUNY Press.

Star, S. L. (ed.) (1995b). *The Cultures of Computing*. Oxford, Blackwell.

Star, S. L., G. Bowker, and L. J. Newman."Transparency at Different Levels of Scale: Convergence between Information Artifacts and Social Worlds." Retrieved 2/23/04 from http://weber.ucsd.edu/~gbowker/converge.html.

Star, S. L. and A. Strauss (1999). "Layers of Silence: Arenas of Voice: The Ecology of Visible and Invisible Work." *Computer Supported Cooperative Work*, 8(1/2): 9–30.

Stephan, N. L. (1993). "Race and Gender: The Role of Analogy in Science." In *The Racial Economy of Science: Toward a Democratic Future*. S. Harding (ed.). Bloomington, Indiana University Press: 359–376.

Strauss, A. (1978). "A Social World Perspective." *Studies in Symbolic Interaction*, 1: 119–128.

Struik, D. (1967). *A Concise History of Mathematics*. New York, Dover.

Tavris, C. (1992). *The Mismeasure of Women*. New York, Simon & Schuster.

Teich, A. (ed.) (2000). *Technology and the Future* (8th edn.). New York, Bedford/St. Martin's Press.

Timmermans, S. (2003). "A Black Technician and Blue Babies." *Social Studies of Science*, 32(2): 197–229.

Tönnies, F. (1987/1963). *Community and Society*. New York, Harper.

Toulmin, S. (1972). "The Historical Background to the Anti-Science Movement. Civilization & Science." In *Conflict or Collaboration?* Ciba Foundation. New York, Elsevier: 23–32.

Traweek, S. (1988). *Beamtimes and Lifetimes: The World of High Energy Physicists*. Cambridge, Harvard University Press.

Tuana, N. (ed.) (1989). *Feminism and Science*. Bloomington, Indiana University Press.

Turkle, S. (1984). *The Second Self: Computers and the Human Spirit*. New York, Simon and Schuster.

Turkle, S. (1995). *Life on the Screen: Identity in the Age of the Internet*. New York, Touchstone.

Turnbull, D. (2000). *Masons, Tricksters, and Cartographers: Comparative Studies in the Sociology of Scientific and Indigenous Knowledge*. Amsterdam, Harwood Academic; Abingdon, Marston.

Tylor, E. (1871/1958). *Primitive Culture*. New York, Harper.

Unger, S. H. (1994). *Controlling Technology: Ethics and the Responsible Engineer*. New York, John Wiley & Sons, Inc.

Valenstein, E. (1998). *Blaming the Brain: The Truth about Drugs and Mental Health*. New York, Free Press.

Valsiner, J. and R. v. der Veer (2000). *The Social Mind: Construction of the Idea*. Cambridge, UK, Cambridge University Press.

Vaughan, D. (1999). "The Dark Side of Organizations: Mistake, Misconduct, and Disaster," *Annual Review of Sociology*, 25: 271–305.

Verran, H. (2001). *Science and an African Logic*. Chicago, IL, University of Chicago Press.

Visvanathan, S. (1997). *A Carnival for Science: Essays on Science, Technology and Development*. Delhi, Oxford University Press.

Volti, R. (ed.) (1995). *Society and Technological Change* (3rd edn.). New York, St. Martin's Press.

Vorzimer, A. W., M. D. O'Hara, and M. D. O'Hara. *Buzzanca v. Buzzanca: The Ruling and Ramifications*, InterNational Council on Infertility Information Dissemination. Retrieved from http://www.vgme.com/buzzanca.html.

Vygotsky, L. S. (1978). *Mind in Society: The Development of Higher Psychological Processes*. Cambridge, Harvard University Press.

Vygotsky, L. S. (1986). *Thought and Language*. Cambridge, MIT Press.

Wagar, W. W. (1967). *The City of Man*. Baltimore, Penguin Books.

Wajcman, J. (1991). *Feminism Confronts Technology*. University Park, Penn State University Press.

Watson-Verran, H. and D. Turnbull (1995). "Science and Other Indigenous Knowledge Systems." In *Handbook of Science and Technology Studies*. Sheila Jasanoff, Gerald E. Markle, James C. Peterson, and Trevor Pinch (eds.). Thousand Oaks, Sage Press: 115–139.

Webster, F. (2003). *Theories of the Information Society* (2nd edn.). New York, Routledge.

Weiner, A. (1976). *Women of Value, Men of Renown: New Perspectives in Trobriand Exchange*. Austin, University of Texas Press.

Weinberg, S. (1992). *Dreams of a Final Theory*. New York, Vintage.

Wertheim, M. (1997). *Pythagoras' Trousers: God, Physics, and the Gender Wars*. New York, W. W. Norton.

Wertsch, J. (1991). *Voices of the Mind: A Socio-cultural Approach to Mediated Action*. Cambridge, Harvard University Press.

Williams, R. (1990). *Notes on the Underground: An Essay on Technology, Society, and the Imagination*. Cambridge, MA: MIT Press.

Wilson, E. (1998). *Neural Geographies: Feminism and the Microstructure of Cognition*. New York, Routledge.

Winner, L. (1977). *Autonomous Technology: Technics-out-of-Control as a Theme in Political Thought*. Cambridge, MA, MIT Press.

Winner, L. (1986). *The Whale and the Reactor: A Search for Limits in an Age of High Technology*. Chicago, University of Chicago Press.

Wittgenstein, L. (1953/2001). *Philosophical Investigations*. Oxford, Blackwell.

Wright, W. (1992). *Wild Knowledge: Science, Language, and Social Life in a Fragile Environment*. Minneapolis, University of Minnesota Press.

Wyer, M., D. Cookmeyer, M. Barbercheck, H. Ozturk, and M. Wayne (eds.) (2001). *Women, Science and Technology: A Reader in Feminist Science Studies*. New York and London, Routledge.

Zazonc, R. B. (1980). "Feeling and Thinking: Preferences Need No Inferences." *American Psychologist*, 35: 151–175.

Zajonc, R. B. (1984). "On the Primacy of Affect." *American Psychologist*, 39: 117–123.

Zenzen, M. and S. Restivo (1982). "The Mysterious Morphology of Immiscible Liquids: A Study of Scientific Practice." *Social Science Information*, 21: 447–473.

Name Index

Abraham, Carolyn, 113
Adas, Michael, 37, 95
Addelson, Kathryn, 3
Adorno, Theodor, 79
Akrich, Madeline, 84
Alexander, 99
Allen, Sally G., 30
Althusser, Louis, 103
Amsterdamska, Olga, viii
Apollonius, 60
Archimedes, 61
Aronowitz, Stanley, 26
Astington, Janet, 43

Bakhtin, Mikhail, 42
Barad, Karen, 40
Barber, Barnard, 11, 118
Barber, Benjamin, 98–100
Barnes, Barry, 25
Baron, J. B., 67
Bartels, J. M., 46
Bartsch, Ingrid, 18
Basalla, George, 85
Beck, Ulrich, 100
Becker, Howard, 4

Benedict, Ruth, 6
Benjamin, Walter, 86
Berger, Peter, 111
Bessel, W., 45
Bibby, C., 118
Bijker, Wiebe, 7, 34, 54, 71, 88
Birke, L., 15
Blake, William, 16
Blakeslee, S., 112
Bleier, Ruth, 15
Blizzard, Deborah, 29
Bloch, Hubert, 15
Bloor, David, 25, 36–40, 116, 119
Boisvert, Yves, 116
Bolyai, J., 45–6
Bolyai, W., 45
Bowker, Geoff, ix
Boyer, Carl, 44–5
Breazeal, Cynthia, 108, 120
Brooks, Rodney, 108, 120
Brothers, Leslie, 43, 109, 114–15,
    120
Butler, Judith, 117
Buzzanca, Jaycee, 104
Byrne, R. W., 42

142

Calvert, M., 101
Campbell, D. T., 109
Casper, Monica, 101
Castells, Manuel, 101
Christ, Jesus, 56, 60, 62–3
Clarke, Adele, 18, 29
Cockburn, Cynthia, 88
Cole, Simon, 47
Collins, Harry, ix, 68, 108
Collins, Randall, 13, 108, 111
Comte, Auguste, 117
Copperfield, David, 55
Cowan, Ruth, 89–90
Crick, Francis, 32
Croissant, Jennifer, 12, 87
Crowley, Aleister, 61

Damasio, A. R., 43
de Montaigne, Michel, 16
Descartes, René, 42
DeSolla Price, Derek, viii
DeVores, B., 43
Dickson, David, 76
Douglas, Mary, 93
Drori, Gili, 94
Durkheim, Emile, xi, 3, 6, 41, 58, 71, 117
Dyens, O., 116

Eberhart, Russell, 109–10
Edison, Thomas, Mrs., 36
Eduardo the Healer, 61
Einstein, Albert, 61, 113, 117
Ellul, Jacques, 73, 76–7, 79, 85, 90
Engels, Frederich, 74
English-Lueck, J. A., 68
Epimethius, 79
Epstein, J., 67
Epstein, Steven, 29, 69
Euclid, 44

Feyerabend, Paul, 14
Finn, J., 86
Fischer, C., 84

Fleck, Ludwig, 48, 65–6
Fodor, J. A., 42
Ford, Henry, 35
Foucault, Michel, 89, 118
Franklin, Rosalind, 32
Fuller, Steve, viii

Galileo Galilei, 61, 117
Gauss, Carl F., 45
Geertz, Clifford, 6, 43
Gerling, A. L., 45
Gibbon, Edward, 60
Gibbs, Lois, 70
Ginsburg, Faye, 101, 120
Ginzberg, Ruth, 30
Goethe, J. Wolfgang, 16
Gordon, S., 43
Gregory the Wonder Worker, 60

Hackett, Edward, viii
Haraway, Donna, 7, 31, 90, 103
Harding, Sandra, 28
Hardt, M., 100
Heelan, Patrick, 109
Heldke, Lisa, 32
Hermes, 79
Hess, David, viii, 25
Hilts, P. H., 112
Hooker, Clifford, 13
hooks, bell, 28
Horkheimer, M., 79
Horton, Robin, 37–40
Hubbs, Joanna, 30
Hughes, Thomas P., 7, 18
Huxley, T. H., 118

Illich, Ivan, 73, 77, 79, 85, 90
Irigaray, L., 29

Jacob, Evelyn, 6
Jasanoff, Sheila, viii, 120
Johnson, Mark, 30

Kant, Immanuel, 42
Kaplan, G., 22
Karp, J., 86
Kastner, Abraham, 45
Kay, L. E., 48
Keller, Evelyn Fox, 30, 31
Kelvin, Lord, 117
Kennedy, James, 109–10
Kepler, Johannes, 94
Klugel, G. S., 45
Knorr-Cetina, Karin, 6
Kreisberg, Seth, 89

Lakoff, George, 30
Lambert, J., 45
Lansing, Steve, 82
Lash, Scott, 94, 119
Latour, Bruno, 6, 7, 25, 111, 116–17, 119
Lave, Jean, 13
Law, John, 72
Lederman, Muriel, 18
Legendre, A.-M., 45
Lemert, Charles, ix
Lenin, V. I., 113
Lobachevsky, N. I., 45–6
Lock, M., 81
Longino, Helen E., 15, 30
Loughlin, Julia, 31, 116
Lynch, Michael, viii

Maddox, B., 32
Makowsky, Michael, 111
Malinowski, Bronislaw, 56
Markle, G. E., viii
Martin, Emily, 15, 25, 30, 93, 108
Marx, Karl, xi, 6, 54, 62, 64, 71, 73–5, 77, 79, 85, 87–9, 90, 117
Marx, Leo, 87
Maxwell, James Clerk, 117
Mead, George Herbert, 3, 42, 109
Merton, Robert K., 11, 51
Mills, C. Wright, 14, 106
Mitchell, R., 120

Mumford, Lewis, 73, 75–6, 79, 85, 88, 90

Nader, Laura, 57, 72
Needham, Joseph, 95
Negri, A., 100
Nelkin, Dorothy, 72
Newman, L. J., 13
Newton, Isaac, 61, 117
Nietzsche, Fredrich, 14, 15, 41–2, 63–6, 71, 108, 115, 118
Nostradamus, 61

O'Keefe, D. L., 57
Olesen, V. L., 18
Ormrod, S., 88
Oudshoorn, Nelly, 48
Overbye, D., 15

Pacey, Arnold, 85, 86, 87
Paterniti, Michael, 114
Perrow, Charles, 92
Pert, Candace, 43
Petersen, J. C., viii
Philip, Kavita, 72
Pinch, Trevor, viii, ix
Pinker, Steven, 112
Plato, 61, 79
Pontius Pilate, 62–3
Pratchett, Terry, 64
Prometheus, 79

Rapp, Rayna, 25, 28, 101, 120
Rappert, Brian, 101
Rasputin, 61
Reich, Robert, 94
Reid, I. S., 29
Reid, R., 18
Reimann, G. F. Bernard, 45, 46
Restivo, Sal, viii, 6, 12, 13, 22, 39, 86, 112, 115, 116, 118, 119
Ritzer, George, 94
Robertson, R., 94, 100
Rochlin, G. I., 83

Rogers, L. J., 22
Rorty, Richard, 28
Rose, Hilary, 26, 28
Rose, Steven, 113

Saccherei, 45
Saint-Simon, 117
Savan, Beth, 70
Schiebinger, Londa, 15, 28
Schiffer, M. B., 16, 36
Schweikart, F. K., 45
Searle, John, 41
Shelley, Mary, 88
Simon Magus, 60
Sinclair, B., 101
Smith, Dorothy, 56, 66, 71, 117
Snell, Pamela, 104
Spengler, Oswald, 3
Spiegel-Rosing, I., viii
Star, S. Leigh, ix, 42, 113
Stephan, Nancy L., 29
Stevin, Simon, 46
Strauss, Anselm, 13
Struik, Dirk, 44–5
Swift, Jonathan, 16

Taurinus, 45
Temple, R., 95
Terry, J., 101
Thomas, Vivien, 8–9
Thurtle, P., 120
Timmermans, Stephen, 7
Toulmin, Stephen, 16
Traweek, Sharon, 6, 18, 30

Tuana, Nancy, 28
Turkle, Sherry, 86–7
Turnbull, David, 101
Tylor, Edward, 6

Unger, Stephen, 7

Valenstein, Edward, 115
Valsiner, J., 43
van der Veer, R., 43
Verran, Helen, 40
Visvanathan, Shiv, 96–7
Vygotsky, L. S., 42, 109

Wajcman, Judy, viii, 85
Watson, James, 32
Weber, Max, xi, 71
Webster, F., 94
Weinberg, Steven, 12
Weiner, Annette, 6
Wenger, E., 13
Wertheim, Margaret, 14
Wertsch, J., 42
Williams, Rosalind, 88
Winner, Langdon, 7, 73, 78–9, 85, 87,
    88, 90–1, 93
Wittgenstein, Ludwig, 2, 41
Woolgar, Steve, 6, 110
Wright, W., 10

Zajonc, R. B., 43
Zeno of Citium, 99
Zenzen, Michael, 6
Zeus, 79

Subject Index

actor-network theory, 26, 85, 91
actor(s), 4, 24, 54, 85–6, 91, 99–100
affordances, 84–5, 89
agency, 75–9, 85, 88–90, 117
    technological agency, xii, 77, 90
AIDS, 27, 29, 69, 93
Alzheimer's, 115
amniocentesis, 28
androcentric, 30, 121
artifact, ix, 2, 7–9, 29, 32, 53, 86, 88,
    90–1, 109, 121
automobile, 33, 35–6, 84
autonomous science, 121
autonomous technology, 78, 80, 88,
    121

bicycle, 33–4
Bill of Rights, 67
brain, 22, 41–3, 66, 81, 112–15

capitalism, xi, 75, 119
class, 3, 8, 20, 25, 28, 35–6, 46, 60, 70,
    73
closure, 34, 121

Colombian fallacy, 13
colonialism, 47, 94, 100
communalism, 51, 121
communities of practice, 3, 7, 54, 115,
    122
consciousness, 41–3, 112–16
convivial society, 77–8, 122
cultural convergence, xii, 81
cultural lag, 105–6
culture, xi, 5–6, 8, 10, 13, 15, 17, 22,
    39, 49–50, 57, 122
    third culture, 22
cyborgs, 90–2, 122

difference, 15, 26–7, 30–2, 88, 94–6
discovery, 4, 6, 22, 29, 32, 45–6, 55,
    61, 98, 117–19
disinterested(ness), 25, 51, 122
domination, 26, 28, 31, 79, 95,
    116

ethical creep, 107, 123
ethnocentric, 44, 123
ethnographies of science, 6, 11

fact(s), xi, 3, 5–6, 8, 11, 13, 16, 19–21, 26, 34, 46, 50, 65–6, 83–4, 108–9, 111, 114, 119
feminism, xi, 26–7, 30–1

gemeinschaft–gesellschaft, 98, 123
gender, xii, 8, 15, 18, 27, 29–31, 46, 70, 73, 87
global economy, 94
glocal, 94, 97, 100, 123
(G)god, 17, 37, 42–3, 55–65, 103, 113, 117
Green Revolution, 82

hard case, 13, 41, 110–14
hegemony, 28, 67, 123

identity, xii, 80, 83, 86–8, 91, 96
indexical, 43
industrial revolution, ix, 77
inequality(ies), 9, 14, 17–18, 27
information technology, 5, 123
innovation, xii, 30–4, 54, 68, 81
inquiry, 4–5, 10–13, 15–17, 25–6, 31, 40, 47, 54, 62–4, 66, 68, 71, 89, 98, 102–3, 115, 117–19
intellectual property, 32
invention, 6, 32–3, 43, 54, 68, 78, 80–1, 83–4, 86–7, 98, 101

knowledge, viii–xii, 5–8, 10–11, 13–14, 16, 18, 22, 28–9, 31–2, 36–7, 39–41, 46, 49, 53–4, 61, 63, 66, 69, 71, 88–9, 93–101, 108, 119
local, 97, 110
production, 30, 37, 70
scientific, 24, 30, 32, 94, 98, 110–11

language, xi–xii, 13, 15, 22, 24, 26, 29–31, 41, 43, 49–52, 58, 93, 97, 104–6
language games, xi, 2, 38
law, xii, 6–7, 11, 35, 49, 67–9, 83, 106
local–global, 99–100

machine, 3, 14, 75–7, 107–9
magic, xii, 23, 49, 54–7, 67, 71
material culture, 49–54, 68
materialism, 124
mathematics, x–xi, 12–14, 41, 44–7, 50, 71, 77, 110–11
megamachine, 75
men, 22, 27, 29, 34, 36, 56, 79, 87, 90
metaphor, 4, 15, 29, 60–1, 93, 97–8, 110, 114
mind, xi–xii, 41–4, 46–7, 66, 86, 107–15
mode of production, 74, 124

narrative, x, 41, 61, 87, 97–8, 101, 103, 110–11, 116–17
nation-state, 22, 26, 96, 100, 103, 124
network(s), 5, 10, 13, 20, 26, 46, 65, 100–1, 114
new reproductive technologies, 28, 101, 104–6, 110
nihilism, 64, 124
nominalism, 124
non-euclidean geometry, 44

objectivity, viii, xii, 2–3, 9, 15, 27, 31, 38, 40, 47, 49, 51, 64, 69, 116–18, 124
organized skepticism, 51, 125

pain, 41
peer review, 22, 53
Pluto, 20–1, 102–3
post-hyphenism, 118–19, 125
postmodernism, 63–4, 66, 71, 91, 103, 116
postmodernist, 97, 116–17, 125
power, ix, xii, 5, 8–10, 13, 16, 24, 26–9, 31, 33–4, 39–40, 60, 63–5, 75, 79, 85–93, 95–7, 100, 108, 119
pre-scientific, 36, 38–40
proof, 5, 13, 38, 46, 61, 100
pure science, 8–9

race, 8–9, 22–3, 29, 32–3, 46, 70, 73, 100
radio, 33
realism, 1, 24–5, 99, 125
reason, 15, 17, 57, 95, 111, 118
reification, 58–9, 125
relativism, 1, 24–5, 47, 64, 102, 111, 116, 125
religion, xii, 17, 26, 28, 49, 51–2, 54–63, 67–8, 71, 76–7, 79, 95, 100, 112
risk analysis, xii, 92–3
ritual, 40, 52–3, 56, 58
robot(s), 103, 107–10

science wars, 1, 126
science/scientific, 5
  discourse, 21
  discovery(ies), 32
  knowledge, 24, 30, 32, 94, 98, 110–11
  method, 5, 21–2, 31, 52, 96
  scientific institution, 15–16, 117
scientist, viii, 1–3, 7, 11–12, 15, 21–5, 27, 31, 38, 43, 49–52, 61–2, 70, 97–8, 100, 102, 106, 112–15, 118
self-organization theory, 126
semiotics, 42, 54, 88, 126
sexism, 27
social, 2, 4, 6, 41, 111
social change, 51, 55, 73–5, 78, 86
social construction, viii, xi–xii, 3, 6, 12, 23–6, 32, 37, 41, 43, 47, 54, 67, 69, 91, 102, 111–12, 117, 119, 126
social construction of technology, 54, 82, 90, 91, 126
social institutions, viii, xii, 4, 5, 10, 22, 35, 40, 49–51, 53, 67, 69, 74, 76, 117, 123
social movements, 31
social networks, x, 6, 20, 46
social processes, 6, 10, 24, 47, 83–4, 110

social relations, xi, 1–2, 8–10, 12, 17, 51, 53, 70–1, 74, 76, 83, 85–6, 90
social sciences, viii, x, 1, 3, 54, 56, 61–3, 117
social solidarity, 41, 52, 58, 63, 84
social theorist, ix, 7, 23, 28, 42, 62, 65, 73, 79
social theory, ix, 6, 28, 41–2, 55–6, 61, 98, 111, 116
social webs of meaning, 102, 126
social world, ix, xi, xii, 4–5, 8, 13–14, 16, 23, 30, 76, 80, 84, 90, 108, 113
society, 6, 22
sociological imagination, 25–6, 66, 114
sociological literacy, 4
standardization, 34, 94
stratification, 18, 50, 55, 57, 60, 127
symbols, 4–5, 14, 19, 29, 49–50, 58–9, 101

technique, 28, 52, 77, 106, 113, 127
technocratic, 96
technology/technological, 7, 74
  change, xii, 4, 10, 54, 78, 80, 81, 83, 92
  culture, 10
  design, 85
  determinism, xii, 10, 78, 80, 83, 100, 127
  diffusion, 32
  fix, 83, 127
  flexibity/affordances, 82, 127
  intensification, 83, 128
  neutrality, 84
  reconstruction, 84
  revolutions, 83
  standards, 81
  systems, x, 54, 83, 91–2
technoscience, viii, xi–xii, 7–12, 15, 49–54, 67–70, 100–1, 102, 128
technosocial, xi–xii, 9, 83, 90–1, 110, 128
theory, 7, 11, 13, 24, 37–8, 40–3, 62, 71, 93, 103–4

tool(s), 1, 3, 5, 9–10, 13–16, 23, 25, 37,
    40, 53–5, 69, 71–2, 74–8, 80,
    89–90, 104
truth, viii–ix, xii, 2, 9, 11, 12, 17, 24–5,
    26, 37, 40, 46–7, 49–50, 62–9,
    103, 111, 116–18
Turing test, 107, 128

Universalism, 28, 47, 51, 128

witchcraft, 38
women, 15, 18, 22, 26–31, 36, 50, 52,
    87, 90, 105–7
worldview(s), xi, 4, 23, 36–9, 44, 46,
    69, 93, 95, 119, 128